Cellular Manufacturing Systems

To Bati Devi, Ravi Kumar and Sheenoo

N.S.

To Sumathi, Prashanthi and Prajan

D.R.

Cellular Manufacturing Systems

Design, planning and control

Nanua Singh
Department of Industrial and Manufacturing Engineering, Wayne State University, Detroit, USA

and

Divakar Rajamani
Department of Mechanical and Industrial Engineering, University of Manitoba, Winnipeg, Canada

CHAPMAN & HALL

London · Glasgow · Weinheim · New York · Tokyo · Melbourne · Madras

Published by Chapman & Hall, 2–6 Boundary Row, London SEI 8HN, UK

Chapman & Hall, 2–6 Boundary Row, London SEI 8HN, UK

Blackie Academic & Professional, Wester Cleddens Road, Bishopbriggs, Glasgow G64 2NZ, UK

Chapman & Hall GmbH, Pappelallee 3, 69469 Weinheim, Germany

Chapman & Hall USA, 115 Fifth Avenue, New York, NY 10003, USA

Chapman & Hall Japan, ITP-Japan, Kyowa Building, 3F, 2-2-1 Hirakawacho, Chiyoda-ku, Tokyo 102, Japan

Chapman & Hall Australia, 102 Dodds Street, Melbourne, Victoria 3205, Australia

Chapman & Hall India, R. Seshadri, 32 Second Main Road, CIT East, Madras 600 035, India

First edition 1996

Softcover reprint of the hardcover 1st edition 1996

Typeset in 10/12 Times by Thomson Press India Limited, Madras

ISBN-13: 978-1-4612-8504-5 e-ISBN-13: 978-1-4613-1187-4
DOI: 10.1007/978-1-4613-1187-4

A catalogue record for this book is available from the British Library

Library of Congress Catalog Card Number: 95-71239

∞ Printed on permanent acid-free text paper, manufactured in accordance with ANSI/NISO Z39.48–1992 and ANSI/NISO Z39.48–1984 (Permanence of Paper).

Contents

Preface

Batch manufacturing is a dominant manufacturing activity in the world, generating a great deal of industrial output. In the coming years, we are going to witness an era of mass customization of products. The major problems in batch manufacturing are a high level of product variety and small manufacturing lot sizes. The product variations present design engineers with the problem of designing many different parts. The decisions made in the design stage significantly affect manufacturing cost, quality and delivery lead times. The impacts of these product variations in manufacturing are high investment in equipment, high tooling costs, complex scheduling and loading, lengthy setup time and costs, excessive scrap and high quality control costs. However, to compete in a global market, it is essential to improve the productivity in small batch manufacturing industries. For this purpose, some innovative methods are needed to reduce product cost, lead time and enhance product quality to help increase market share and profitability. What is also needed is a higher level of integration of the design and manufacturing activities in a company. Group technology provides such a link between design and manufacturing. The adoption of group technology concepts, which allow for small batch production to gain economic advantages similar to mass production while retaining the flexibility of job shop methods, will help address some of the problems.

The group technology (GT) approach originally proposed by Mitrofanov and Burbidge is a philosophy that exploits the proximity among the attributes of given objects. Cellular manufacturing (CM) is an application of GT in manufacturing. CM involves processing a collection of similar parts (part families) on a dedicated cluster of machines or manufacturing processes (cells). The cell formation problem in cellular manufacturing systems (commonly understood as the cell design problem in literature) is the decomposition of the manufacturing systems into cells. Part families are identified such that they are fully processed within a cell. The cells are formed to capture the inherent advantages of GT like reduced setup times, reduced in-process inventories, improved product quality, shorter lead time, reduced tool requirements, improved productivity, better overall control of operations, etc. The common disadvantages are lower machine and labor utilization and higher investment due to duplication of machines and tools.

The problem of cell design is a very complex exercise with wide ranging implications for any organisation. Normally, cell design is understood as the problem of identifying a set of part types that are suitable for manufacture on a group of machines. However, there are a number of other strategic level issues such as level of machine flexibility, cell layout, type of material handling equipment, types and number of tools and fixtures, etc. that should be considered as part of the cell design problem. Further, any meaningful cell design must be compatible with the tactical/operational goals such as high production rate, low WIP, low queue length at each work station, high machine utilization, etc. A lot of research has been reported on various aspects of design, planning and control of cellular manufacturing systems. Various approaches used include coding and classifications, machine-component group analysis, similarity coefficients, knowledge-based, mathematical programming, fuzzy clustering, clustering, neural networks, and heuristics among others.

The emphasis in this book is on providing a comprehensive treatment of various aspects of design, planning and control of cellular manufacturing systems. A thorough understanding of the cell formation problem is provided and most of the approaches used to form cells are provided in Chapters 2 through 7. Issues related to layout design, production planning and control in cellular manufacturing systems are covered in Chapters 8, 9 and 10 respectively.

The book is directed towards first and second year graduate students from the departments of Industrial, Manufacturing Engineering and Management. Students pursuing research in cellular manufacturing systems will find this book very useful in understanding various aspects of cell design, planning and control. Besides graduate engineering and management students, this book will also be useful to engineers and managers from a variety of manufacturing companies for them to understand many of the modern cell design, planning and control issues through solved examples and illustrations.

The book has gone through thorough classroom testing. A large number of students and professors have contributed to this book in many ways. The names of Dr G.K. Adil, Pradeep Narayanswamy, Parveen S. Goel and Saleh Alqahtany deserve special mention. We are grateful to Dr Jeffery S. Smith of Texas A & M University and Dr Sanjay Joshi of Penn State University for contributing Chapter 10 in this book.

We are also indebted to Mark Hammond of Chapman & Hall (UK) for requesting us to write this book. We appreciate his patience and tolerance during the preparation of the manuscript.

The cover illustration is reproduced courtesy of Giddings & Lewis (USA).

Nanua Singh
Divakar Rajamani

September 1995.

CHAPTER ONE

Introduction to design, planning and control of cellular manufacturing systems

The long-term goals of a manufacturing enterprise are to stay in business, grow and make profits. To achieve these goals it is necessary for these enterprises to understand the business environment. The twenty-first century business environment can be characterized by expanding global competition and customer individualism leading to high-variety products which are low in demand. In the 1970s the cost of products used to be the main lever for obtaining competitive advantage. In the 1980s quality superseded cost and became an important competitive dimension. Now low unit-cost and high quality products no longer solely define the competitive advantage for most manufacturing enterprises. Today, the customer takes both minimum cost and high quality for granted. Factors such as delivery performance, customization of products and environmental issues such as waste generation are assuming a predominant role in defining the success of manufacturing enterprises in terms of increased market share and profitability. The question is: what can be done under these changing circumstances to stay in business and retain competitive advantage?

What is needed is the right manufacturing strategy to meet the challenges of today's and future markets. In doing so, a manufacturing organization not only has to understand what customers want, it also has to develop internal mechanisms to respond to the changes demanded by what the customer wants. This requires a paradigm shift in everything that factories do. That means making use of state-of-the-art technologies and concepts. From a customer view point, a company has to respond to smaller and smaller market niches very quickly with products that will get built in lower and lower volume at the minimum possible cost. The concepts of group technology/cellular manufacturing can be utilized in such a high variety/low demand environment to derive the economic advantages inherent in a low variety/high demand environment. This book provides a comprehensive treatment of various issues on the design, planning and control of cellular manufacturing systems.

1.1 PRODUCTION SYSTEMS AND GROUP TECHNOLOGY

Modern industrial production differs in the nature and use of 'different' equipment, end-products and type of industry. While these are the bases for differences as far as the management of production activities is concerned, it is primarily the size of the production volume in relation to cost and delivery promises which imposes problems. Thus, the nature of production processes can be classified as intermittent, continuous or repetitive.

When the parts (jobs) that arrive at the job shop are different from one another and the demand is intermittent (for example an auto repair shop), it is suitable to have a standard machine layout to cater for all varieties. A 'job shop layout' (process layout) is best suited to low volume and high variety. A typical job shop has several departments, where each department provides processing facilities for specific operations, for example, drilling, and milling. The parts have to move from one department to another for various operations. The planning, routing and scheduling function has to be done for each part independently. The difference in product design requires versatile equipment, thus making specialized equipment uneconomical. With such a layout, the part spends a substantial amout of time (about 95%) waiting before and after processing, on traveling between departments and on setup. The time lost on wait, travel and setup increases the manufacturing lead time resulting in low productivity.

In contrast, when a shop is engaged in large-scale production of only a few part types, it is possible to arrange the machines in a sequence such that a continuous flow is maintained from start to finish. Specialized equipment is affordable due to production volumes in this case. This type of layout is referred to as the 'flow shop layout' (product layout). After the layout has been arranged and the workstations balanced, the problem of routing and scheduling can be done with the stroke of a pen. The cost of production is lowest in this type of layout. Conversely, interruptions become extremely costly. Also, changing the layout for the production of a different item is costly in terms of lost production.

Between these two extremes a number of enterprises deal with a repetitive demand. This is often referred to as 'batch production'. The extent of overlap with intermittent and continuous is often vague. A number of process industries can use the same production line for mixing a variety of substances, such as chemicals, soups, toothpaste etc. Thus, in contrast to a pure flow line, the plant built for repetitive operations produces different types of products. Certain job shops may also prefer to operate on a repetitive basis if a sizeable order exists. A 'combination layout' is usually proposed for this wide zone of operations. So far as control is concerned, repetitive production bears similarities with both intermittent and continuous production. Thus, this is also characterized by high setup times, high lead times and low productivity.

The concept of **group technology** (GT) has emerged to reduce setups, batch sizes and travel distances. In essence, GT tries to retain the flexibility of a job shop with the high productivity of a flow shop. It originally emerged as a single-machine concept as created by Mitrofanov in Russia. A number of similar parts were grouped and loaded successively on to a machine in order to maximize use of a single setup, or to reduce the setting necessary to produce the group of parts. Thus, machine utilization (i.e. actual operating times) could be increased above the 40% level accepted as normal in a functional layout-based system (Jackson, 1978). This grouping allows the use of high-output machines, which were previously uneconomical due to large setup times in a job shop layout. This concept was further extended by collecting parts with similar machining requirements and processing them completely within a machine group (cell). Depending on the process requirement and the sequence and variety of parts, the flow of material within a cell could be jumbled or continuous (the trend is towards a 'U' flow). Thus, depending on the production volume, the GT manufacturing system can involve the following three forms (Arn, 1975): the GT center, the GT cell and the GT flow line. These three layouts lie between the job shop layout for small batch production and flow shop layout as representative of large batch production. Details on the characteristics of these layouts are given in Chapter 8. These cellular manufacturing systems could be manned as well as unmanned. The unmanned systems are often referred to as **cellular flexible manufacturing systems**. Also, in some cases a few machines or processes are immovable and are required by a large variety of parts. These can be referred to as shared resources and placed in a cell called as the **remainder cell** or each of these could be a GT center.

Besides the principal difference with respect to similarity or dissimilarity between parts, the variety and volume of parts, there are other aspects of cellular manufacturing systems which are important. These include process planning, production planning and machine loading, which will differ depending on the cell structure. Greene and Sadowski (1984) defined the complexity of the GT system by the terms 'machine density' and 'job density'. Machine density refers to the commonality of machine types between cells. The machine density depends on the cell characteristics, which depend on the number of cells, the number of machine types per cell, the total number of different machine types and the remainder cell. In contrast, the job density is defined as the proportion of cells that jobs could be feasibly assigned to. Job density encompasses job characteristics such as the number of operations per job and the number of job types. It also includes the cell characteristics.

Formally, GT refers to the logical arrangement and sequence of all facets of company operations in order to bring the benefits of mass production to high variety, medium-to-low quantity production (Ranson, 1972). Thus it can be said that GT is a change in management philosophy. The application of GT to manufacturing called cellular manufacturing is a manufacturing strategy to win a war against global competition by reducing manufacturing costs, improving quality and by reducing the delivery lead time of products in a high variety, low demand environment.

The following sections briefly review the impact of GT and cellular manufacturing on system performance, and on other functional areas and technologies.

1.2 IMPACT OF GROUP TECHNOLOGY ON SYSTEM PERFORMANCE

The benefits derived from the GT manufacturing system in comparison with the traditional system in terms of system performance will be discussed in this section.

Material handling

In GT layout, the part is completely processed within a cell. Thus, the part travel time and distance is minimal. In a local industry it was found that two products consisting of 11 subassemblies traveled a total of 64 km. In contrast, if the machines are placed in a cell, the total distance traveled will be approximately 6 km, an improvement of about 10 times. This benefit, however, depends on the existing layout shape and size. If the distances are not appreciable, GT can be practised without physically moving machines, but rather by identifying cliques and dedicating them to a collection of parts.

Throughput time

In a traditional job shop, a part moves between different machines in a batch which is often the economic batch size. For example, consider a part of batch size 100, which requires three operations and each operation takes 3 min. Assuming negligible travel times, the batch is completed after 900 min (100 × 3 × 3). The same part if routed through a cell consisting of the three machines will take 9 min for the first part, with each subsequent part produced after every 3 min, thus taking a total of 306 min (99 × 3 + 9). This represents an improvement of about three times. This improvement is feasible because of the proximity of machines in the cell, thus allowing the production control to produce parts as in a flow shop.

Setup time

The reduction in setup time is not automatic. In GT, since similar parts have been grouped, it is possible to design fixtures which can be used for a variety of parts and with minor change can accommodate special parts. These parts should also require similar tooling, which further reduces the setup time. In the press shop at Toyota, for example, workers routinely change dies in presses in 3–5 min. The same job at GM or Ford may take 4–5 h (Black, 1983). The development of flexible manufacturing systems further contributes to the reduction in setup by providing automatic tool changers and also a reduction in processing time, producing high-quality products at low cost.

Batch size

Due to high variety and setup times, a job usually produces parts based on the economic batch quantity (EBQ). This is a predetermined quantity which considers the setup cost (fixed cost, independent of quantity) and the labor, inventory and material costs (variable cost, depends on quantity). The fixed cost must be distributed over a number of parts to make production economical. As the quantity increases the inventory cost increases. In GT, however, the setup can be greatly reduced, thus making small lots economical. Small lots also smooth the production flow, an ideal lot size being one. This in principle is the philosophy of just-in-time production systems, and GT in essence becomes a prerequisite.

Work-in-progress

In a job shop, the economic order quantity for different parts at different machines varies due to differences in setup and inventory costs. The different consumption rates of these parts in the assembly will inevitably lead to shortages. Rescheduling the machines to account for these shortages will increase setup cost and provide additional safety stock for other parts. The delivery times and throughput times are fuzzy in this situation. A level of stocks equal to 50% of annual sales is not unusual, and is synonymous with batch production (Ranson, 1972). GT will provide low work-in-progress and stocks. This is also due to the type of production control and will be discussed in a later section.

Delivery time

The capability of the cell to produce a part type at a certain predetermined rate makes delivery times more accurate and reliable.

Machine utilization

Due to the decrease in setup times the effective capacity of the machine has increased, thus leading to lower utilization. This is working smart and short, not a disadvantage as is often stated. Also, to ensure that parts are completely processed within a cell, a few machines have to be duplicated. These machines are of relatively low cost and are the ones often underutilized. By changing process plans and applying value engineering, it is possible to avoid this investment by routing these operations on existing machines which now have more capacity. The general level of utilization of cells (except the key machines) is of the order of 60–70%. In a job shop, the primary objective of the supervisor and management is to use the machine to the fullest. If any machines are idle, parts are retrieved from stores and the EBQ for that part is processed on these machines to keep the machines busy. This is essentially adding value to parts, stacking inventory, and is another manifestation of making unwanted things. With current market trends many of these items will be obsolete before they leave the factory. Hollier and Corlett (1966), after studying a number of companies involved in batch production, concluded that undue emphasis on high machine utilization only results in excessive work-in-progress and long throughput times.

Investment

To ensure parts are completely processed in a cell a few machines have to be duplicated. Often these machines are of relatively low cost. The major investment would be the relocation of machines and the cost of lost production and reorganization. However, this cost is easily recovered from the inventory, better utilization of machines, labor, quality, material handling etc.

Labor

Due to lower utilization levels of the cell, it is possible to have better utilization of the workforce by assigning more than one machine to a worker. This leads to job enrichment, and with rotations within a cell, these people form a team whose objective is to produce a complete product which gives them job satisfaction. There is considerable evidence that a team working together will produce more than an individual. This also forms the basis for total quality management.

Quality

Since parts travel from one station to another as single units (or small batches), the parts are completely processed within a small region, the feedback is immediate and the process can be halted to find what went wrong.

Space

Due to the decrease in work-in-progress, there will be considerable floor space available for adding machines and for expansion.

1.3 IMPACT ON OTHER FUNCTIONAL AREAS

This section discusses the influence of GT on different functional areas in a manufacturing enterprise.

Part design

Part proliferation due to the absence of design standards is common among discrete-part manufacturers. The cost of proliferation affects not only the design area, but also the release of parts to manufacturing. The expenses of release to manufacturing include charges for part design, prototype building, testing and experimentation, costing, records and documentation. One source estimates these costs range between $1300 and $2500 per part. Others have indicated that the figures can vary between a low of $2000 and a high of $12 000 (Hyer and Wemmerlov, 1984). The application of GT assists in the identification of similar parts, thus reducing the variety, promoting standardization and decreasing the number of new part designs.

Production control

The traditional control procedure is 'stock control' in which each part is manufactured in EBQs and held in stock. When withdrawals cause the stock to fall below a predetermined point, a new batch is manufactured. The disadvantage of this has already been emphasized. GT has the characteristics which makes it more suitable for flow control such as in flow layouts. Thus the manufacture of parts is related directly to short-term demand, and variable batch sizes are produced at a fixed frequency. Although this may be suitable for most parts, for some low value/volume items where the demand is unpredictable, a stock control policy may be adopted. With all the parts assigned to specific cells, the production control function is relatively simple in comparison to a job shop. Simple control boards are sufficient to determine the loading sequence.

Process planning

Computer-aided process planning is an essential step towards computer integrated manufacturing. The largest productivity gains due to GT have been reported in this area. With GT-coding, it is possible to standardize such plans, reduce the number of new ones, and retrieve and print them out efficiently (Hyer and Wemmerlov, 1984)

Maintenance

With GT, a preventive maintenance program becomes essential. Since each machine is dedicated to a part family, the flexibility to re-route these parts on similar machines does not exist. Thus, as with flow lines, the cost of downtime is high. However, with proper training, operators can perform regular maintenance. This leads to improved machine life, job enrichment and group responsibility to maintain the machines.

Accounting

Each cell is now a cost center. Since the complete part is produced within the cell, costing is easier. Moreover, depending on the similarity of parts within a family, the cost structure for the parts can be easily established. When a part is not processed on all machines, this should be accounted for, and it is easy to do so. More accurate costing information can be obtained considering the age, performance and investment on machines within a cell. In contrast, in a job shop the part could be processed on one of a number of similar machines for which these factors are different.

Purchasing

GT can help reduce the proliferation of purchases. An aerospace group which produced engine nuts purchased blank slugs based on part number demand. By using a GT-coding system the company found that fewer different parts could be purchased in higher volumes, resulting in an annual saving of $96 000 (Hyer and Wemmerlov, 1984).

Sales

The basic principles of industrial engineering are standardization, simplification and specialization. The success of GT depends on adopting this concept not only in the design of new products but also by the sales department, where they sell by current designs. If specialized items are required, they should be carefully considered within the boundary of company objectives. With a GT system, since each cell is a cost center, more accurate costing and delivery times can be quoted by sales to the customers.

1.4 IMPACT ON OTHER TECHNOLOGIES

The impact of GT on a number of philosophies/technologies will be discussed in this section.

Numerical control (NC) machines

As stated earlier, GT assists in the economic justification of expensive NC machines in a job shop.

Flexible manufacturing systems

The need for flexibility and high productivity has led to the development of flexible manufacturing systems (FMSs). A FMS is an automated manufacturing system designed for small batch and high variety of parts which aims at achieving the efficiency of automated mass production while maintaining the flexibility of a job shop. The benefits of GT stress that FMS justifications must proceed on the basis of explicit recognition of the nature of the GT system.

Computer integrated manufacturing

The progression from the functional shop to manned cells to clusters of CNC machines to an entire system of linked cells must be accomplished in logical, economically-justified steps, each building from the previous state (Black, 1983). GT paves the way for this progression.

Material requirements planning (MRP)

This is a production control system where ordering is based on EBQs. The recommended system for control in a GT system is 'period batch control' (PBC). PBC bases part ordering on new sales and production programs. The quantities of products scheduled for assembly are the just-in-time quantities plus, in some conditions, occasional additions for smoothing. The time cover, or term, of these programs varies from 2 to 5 weeks. The order quantity for parts manufactured for each period is the same as the requirement quantity for the following period. Dispatching, including operation scheduling, is normally delegated to the GT group. Although it can be treated as a special variant of MRP, there is at present no recorded application with these characteristics (Burbidge, 1989).

Just-in-time

Continuous reductions of setup time, lead time and inventory which achieve a streamlined production process through merging processes by bringing machines to the 'point of use' are also the major mandates of the just-in-time (JIT) philosophy. As one machine cannot join several different GT systems, the JIT production system calls for rethinking of the way plant is equipped with machines. Conventional wisdom, when a machine is needed, is to go for the 'big, fast one'. Such 'wisdom' may be unwise (Schonberger, 1983). Success stories speak of small, slow, cheap machines modified to fit the cell. Machine utilization is of little importance. Having the machine at the point of use, dependable and ready to go, is important for JIT. Although GT can be practised without JIT, it is a prerequisite for JIT.

Concurrent engineering

Steady and predictable demand are desirable for a GT manufacturing system. However, depending on the nature of the customer, this may not be a controllable factor. Moreover, machine rearrangements on a short-term basis are not economical. Concurrent engineering provides a way to increase the life of a cell. It brings the product design and process design functions together. Thus, one is able to identify the cell capabilities and ensure, as far as possible, developing a design which can be produced using current capabilities.

1.5 DESIGN, PLANNING AND CONTROL ISSUES IN CELLULAR MANUFACTURING

Group technology provides a means to identify and exploit similarities of products and processes. In product design the focus of GT is on geometric similarities. Part families with similar functions, shapes and sizes can be formed. When a new part is to be designed, the designer can use the database for existing part families which are similar in functionality and geometric features. This reduces engineering design time and design cost. In manufacturing engineering, similarities in machining operations, tooling and setup procedures, transporting and storing materials are exploited. Parts can be grouped into families based on these similarities. Processing parts together in these dedicated cells leads to most of the benefits outlined in section 1.3. Various cell formation approaches are given in Chapters 2 to 7. However, to realize fully the benefits of group technology and cellular manufacturing, it is important to understand and characterize the post-design issues such as planning and control. This book provides a comprehensive coverage of the approaches used in the design, planning and control of cellular manufacturing systems. In Chapter 8, a comprehensive treatment of layout planning in cellular manufacturing is given. Production planning issues are discussed in detail in Chapter 9. A detailed treatment of control issues in cellular flexible manufacturing systems is given in Chapter 10.

1.6 OVERVIEW OF THE BOOK

The focus of the book is on modeling and analysis of cellular manufacturing systems. A detailed analysis is provided of various models and solution procedures used in the design, planning and control of cellular manufacturing systems. Accordingly, the material in the book is organized in a logical order following the design, planning and control phases of manufacturing in a cellular manufacturing environment.

The primary advantage of a GT implementation is that a large manufacturing system used to produce a set of parts can be decomposed into smaller subsystems of part families based on similarities in design attributes and manufacturing features. A number of GT approaches have been developed to decompose a large manufacturing system into smaller, manageable systems based on these similarities. A classification approach using coding systems is one such approach. Chapter 2 provides a detailed discussion of various coding systems. A number of clustering algorithms, such as the hierarchical clustering algorithm, the P-median model and multi-objective clustering, are described and illustrated by numerical examples.

To derive economic advantages, parts can be divided into part families and existing machines into associated groups by analysing the information in the process routes. The part–machine group analysis methods essentially use the information obtained from the routing cards which is represented in a matrix called the 'part–machine matrix'. The part–machine matrix has 0–1 entries, 1 signifying that an operation on a machine is done and 0 signifying that an operation is not done. A number of algorithms have been developed which exploit this information to form cells. In Chapter 3 several of these algorithms are presented, including the bond energy algorithm, rank order clustering

(ROC), ROC2, modified ROC, the direct clustering algorithm, the cluster identification algorithm (CIA) and the modified CIA. A number of performance measures are discussed and compared.

In Chapter 4 the concepts of similarity coefficients are introduced. Based on the similarity coefficients, a number of algorithms are presented to form cells. These algorithms include single linkage clustering, complete linkage clustering and linear cell clustering. The concepts of machine chaining and of single machines are discussed. A discussion on the procedures for cell evaluation and groupability of data is also given.

The above cell formation approaches are essentially heuristics. The notion of an optimal solution provides a basis for comparison as well as an understanding of the structural properties of the cell design problems. A number of mathematical models are provided in Chapter 5, which include the p-median, assignment, quadratic programming, graph theoretic and nonlinear programming models. Various algorithms are compared and the results are discussed.

The primary objective of Chapter 6 is to provide applications of simulated annealing, genetic algorithms and neural networks in cell design. These techniques are becoming popular in combinatorial optimization problems. Since the cell design problem is combinatorial in nature, these techniques are successful in cell design.

In Chapter 7, a number of manufacturing realities are considered in cell design. For example, the problem of alternative process plans is introduced. The models for sequential and simultaneous formation of cells are presented. The cost of material handling and relocation of machines are considered in designing new cells. Further, we provide a cell formation approach which considers the trade-offs between the setup costs and investment in machines.

Layout planning in any manufacturing situation is a strategic decision. At the same time it is a complex problem. Volumes of literature have been written and a number of packages have been developed. Chapter 8 focuses only on the cellular manufacturing situation. Accordingly, the discussion is limited to various types of GT layouts. Some mathematical models are presented for single- and double-row layouts typical of cellular manufacturing systems.

One of the most important aspects of cellular manufacturing is cell design. Once the cells have been designed and machines have been laid out in each cell, the next obvious issue that should be addressed is production planning. The allocation of operations on each machine has to be addressed. The allocations differ based on criteria such as minimizing unit processing cost, or minimizing total production time or balancing workloads. Chapter 9 provides a basic framework for production planning which exploits the benefits of both MRP as well as

GT. We also provide mathematical models which take advantage of group setup times.

Once the machines have been laid out and operations assignments have been made, decisions on the sequencing of parts and the operations on these parts as well as cooperation among various machines, robots and other material handling equipment are important issues in optimizing the performance of cellular manufacturing systems. Various shopfloor control architectures are defined in Chapter 10, and a hierarchical control architecture is discussed in detail. The use of state tables and Petri nets for implementing shopfloor control are described in detail.

1.7 SUMMARY

Group technology is a management strategy. It affects all areas of a company and its impact on productivity cannot be underestimated. To implement a GT system of production successfully, one has to understand its impact on the system performance, the functioning of different departments and the technologies that assist in the implementation. If it is introduced well, it can lead to economic benefits and job satisfaction. This chapter introduced the concepts of GT and cellular manufacturing. A chapter-by-chapter scheme detailing the design, planning and control issues in cellular manufacturing was provided.

PROBLEMS

1.1 What is group technology? Discuss in brief its application to manufacturing.
1.2 What are the advantages and disadvantages of the 'single machine concept'?
1.3 Compare the GT system of manufacturing with traditional systems.
1.4 Discuss the impact of GT on machine utilization. Does it have a negative impact?
1.5 How does GT assist in the implementation of the just-in-time system?
1.6 Discuss the importance of considering the system design and system operation parameters simultaneously during cell design.

REFERENCES

Arn, E.A. (1975) *Group Technology: an Integrated Planning and Implementation Concept for Small and Medium Batch Production*, Springer-Verlag, Berlin.

Black, J.T. (1983) Cellular manufacturing systems reduce setup time, make small lot production economical. *Industrial Engineering*, November, 36–48.

Burbidge, J.L. (1989) A synthesis for success. *Manufacturing Engineer*, November, 29–32.

Greene, T.J. and Sadowski, R.P. (1984) A review of cellular manufacturing assumptions, advantages and design techniques. *Journal of Operations Management*, **4**(2), 85–97.

Hollier, R. and Corlett, N. (1966) *Workflow in Batch Manufacturing*, HMSO, London.

Hyer, N.L. and Wemmerlov, U. (1984) Group technology and productivity. *Harvard Business Review*, **62**(4), 140–49.

Jackson, D. (1978) *Cell System of Production: an Effective Organization Structure*, Business Books, London.

Ranson, G.M. (1972) *Group Technology: a Foundation for Better Total Company Operation*, McGraw-Hill, London.

Schonberger, R.J. (1983) Plant layout becomes product-oriented with cellular, just-in time production concepts. *Industrial Engineering*, November, 66–71.

FURTHER READING

Burbidge, J.L. (1991) Production flow analysis for planning group technology. *Journal of Operations Management*, **10**(1), 5–27.

Gallagher, C.C. and Knight, W.A. (1986) *Group Technology: Production Methods in Manufacture*, Ellis Horwood, Chichester.

Ham, I., Hitomi, K. and Yoshida, T. (1985) *Group Technology: Applications to Production Management*, Kluwer-Nijhoff Publishing, Boston.

Singh, N. (1993) Design of cellular manufacturing systems: an invited review. *European Journal of Operational Research*, **69**, 284–91.

Vakharia, A.J. (1986) Methods of cell formation in group technology: a framework for evaluation. *Journal of Operations Management*, **6**(3), 257–71.

CHAPTER TWO

Part family formation: coding and classification systems

Batch manufacturing produces a variety of different parts and accounts for 60–80% of all manufacturing activities (Chevalier, 1984). Moreover, at least 75% of all such parts are made in batches of less than 50 units (Groover, 1987). This large variety of parts and small batch sizes leads to part design and manufacturing inefficiencies such as inefficient use of design data, inaccuracies in planning and cost estimation, poor workflow, high tooling cost, high setup cost, large inventories and delivery problems. The remedy to these problems lay in sorting parts into families that have similar part design attributes and/or manufacturing attributes for a specific purpose. Design attributes include part shape (round or prismatic), size (length/diameter ratios), surface integrity (roughness, tolerance etc.), material type, raw material state (casting, bar stock etc.) etc. The part manufacturing attributes include operations (turning, milling etc.) and sequences, batch size, machine and cutting tools, processing times, production volumes etc.

The purpose of the family determines the attributes to be considered. For example, if part design advantages are to be gained then parts of identical shape, size etc. which are based on design attributes are in one family. This allows design engineers to retrieve existing drawings to support new parts. Further, when these attributes are standardized it prevents part variety proliferation, and provides accurate planning and cost estimation values. For the purpose of manufacturing, however, two parts identical in shape, size etc. may be manufactured in different ways and hence may be members of different families. Manufacturing efficiencies are gained from reduced setup times, part family scheduling, improved process control, standardized process plans, improved flow etc. 'Part family formation' thus takes advantage of similarities between parts and increases effectiveness by (Hyer and Wemmerlov, 1984):

- performing like activities together;
- standardizing similar tasks;

- efficiently storing and retrieving information about recurring problems.

An engineering database containing information on part design and manufacturing attributes also provides a bridge between computer-aided design and manufacturing (Billo, Rucker and Shunk, 1987).

Part family formation is, therefore, a prerequisite for the efficient manufacture of parts in groups and is probably the main determinant for the overall effectiveness of the cell system of production. The original approach as used in Russia was to divide the total range of parts according to similarity of equipment (lathe, milling, drill etc.) required for manufacture; then by geometric shape (shafts, bushes etc.); thirdly by design type (rings, mounts, gears etc.) and finally by similarity of tooling equipment, as shown in Fig. 2.1. This process identified similar parts and led to the development of the composite part concept. A composite part is a complex part which incorporates all, or most, of the

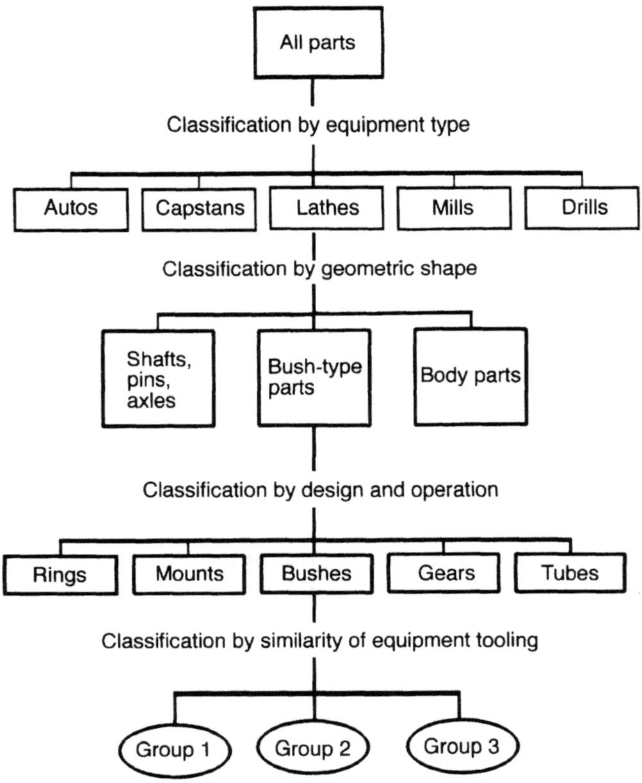

Fig. 2.1 Original approach to part family formation Jackson (1978).

design features of a family of similar parts. This theoretical composite part is extremely useful in the development of tooling layouts on machines (Jackson, 1978). This approach was used to load single machines, but when part families and machine groups were formed by considering sequential operations, it suffered from low utilization of the secondary operation machines.

In cases where the part variety is low a visual/manual analysis by part and drawing can be used to determine part families. When the part variety is large, to consider all factors it is preferable to code all the parts and classify parts by the code similarity or distance. 'Cluster analysis' is a generic name for a variety of mathematical methods, numbering hundreds, that can be used to find parts which are similar or distant from one another. This chapter uses three commonly used distance measures to distinguish between parts. Part families will be identified using one of the following clustering algorithms: the hierarchical clustering algorithm, the p-median model or the multi-objective clustering algorithm.

2.1 CODING SYSTEMS

A code is a string of characters which store information about a part. Using a coding system all the digits are assigned numerical codes (all numbers), alphabetical codes (all letters) or alphanumeric codes (mixed numbers and letters). Depending on how the digits of a code are linked, there are three coding systems: monocode (hierarchical code), polycode (attribute code) and mixed code.

Monocode

The system was originally developed for biological classification, where each digit code is dependent on the meaning of the previous code. An example of the tree structure thus developed is shown in Fig. 2.2. Some of the main features of this scheme are that it:

- is difficult to construct;
- provides a deep analysis;
- is preferred for storing permanent information (part attributes rather than manufacturing attributes);
- captures more information in a short code (fewer digits needed);
- is preferred by design departments.

Polycode

Each digit in a polycode describes a unique property of the part and is independent of all the other digits. An example of a polycode is shown

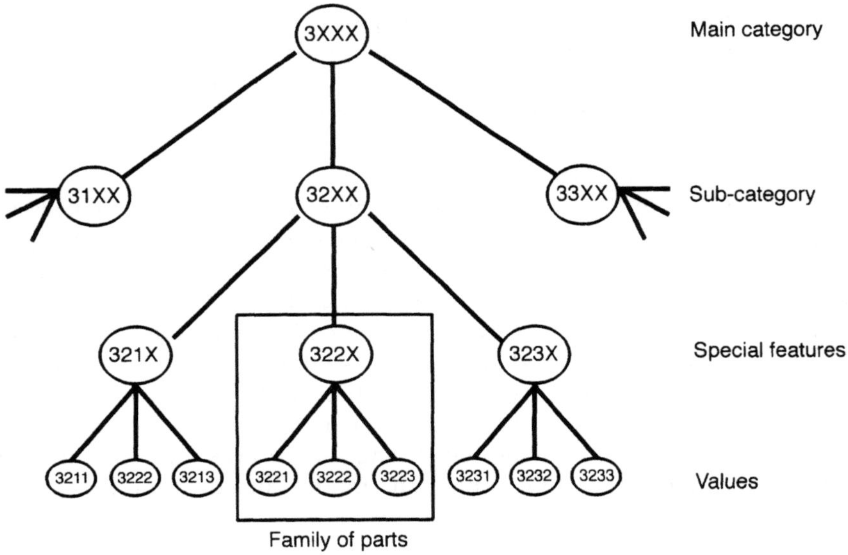

Fig. 2.2 Monocode system of classification. (Printed with permission from *American Machinist*, December 1975. A Penton Publication.)

in Fig. 2.3. The main features of this scheme are:

- it is easy to learn, use and alter;
- it is preferred for storing impermanent information (manufacturing features);
- the length of code may become excessive because of its limited combinatorial features;
- it is preferred by manufacturing departments.

Mixed code

To increase the storage capacity, mixed codes consisting of few digits connected as monocode followed by the rest of the digits as polycode

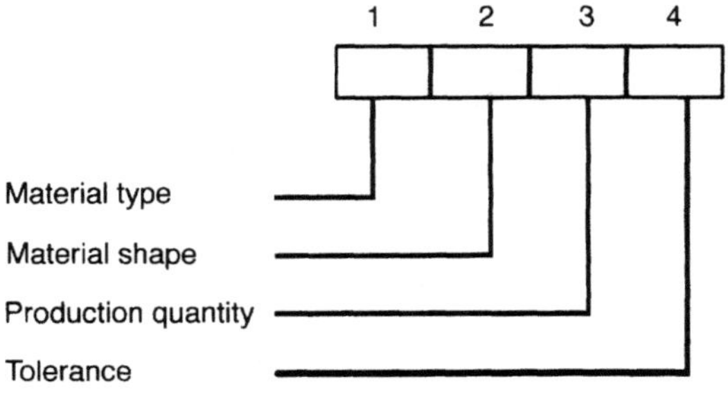

Fig. 2.3 Polycode system of classification.

are usually preferred. The benefits of both systems are thus combined in one code.

2.2 PART FAMILY FORMATION

Classification or part family formation is the process of grouping similar parts or separating dissimilar parts based on predetermined attributes. For example, the parts may be classified on the basis of geometric shapes, dimensions, type of material, operation etc. Codes are a vehicle through which this identification takes place. If a monocode is used, a family is defined as a collection of 'end twigs' and their common node (see Fig. 2.2). To increase the family size, we would have to go up to the next node and include all branches attached at that point (Fig. 2.4). This process can become cumbersome. The situation with mixed code is even worse (Eckert, 1975).

In the following sections, emphasis is on forming part families where polycodes are used. Part families determined by considering manufacturing attributes and design attributes are not necessarily connected. Therefore, it is of utmost importance to define and compare only the necessary attributes. Thus we want '*f*' part families identified from a set of *P* parts for the desired objective function. To define the objective function, three commonly used distance measures are defined next.

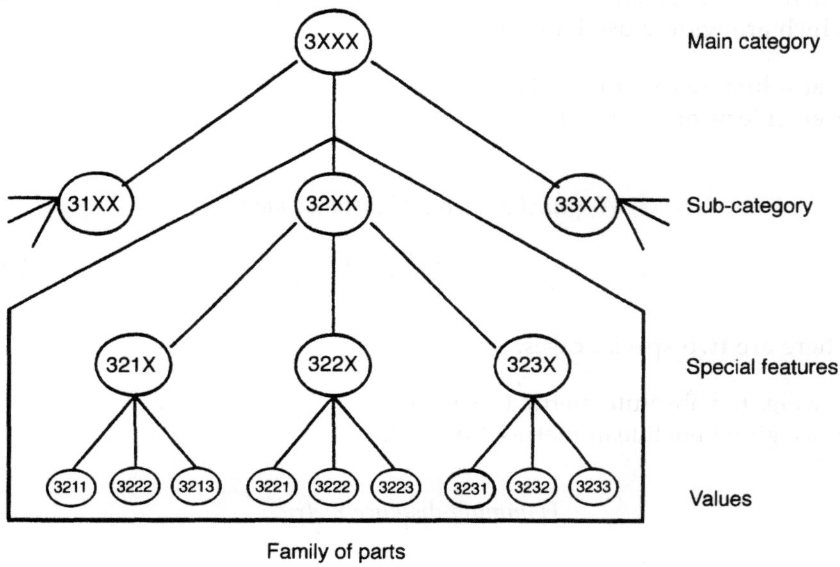

Fig. 2.4 Monocode system of classification. (Printed with permission from *American Machinist*, December 1975. A Penton Publication.)

Distance measures

Each part p can be assigned a vector $\mathbf{X}_p$ of attribute values (Han and Ham, 1986)

$$\mathbf{X}_p = [\mathbf{X}_{pk}]$$

where $\mathbf{X}_{pk}$ is the kth attribute value of part p, K is the number of digits of the coding system, and $k = 1$ to K. For two codes $\mathbf{X}_p$ and $\mathbf{X}_q$ for parts p and q, a distance d_{pq} can be defined which is a real-valued symmetric function, obeying the three axioms (Fu, 1980)

- reflexivity, $d_{pp} = 0$
- symmetry, $d_{pq} = d_{qp}$
- triangle inequality, $d_{pq} \leqslant d_{ps} + d_{sq}$

where s is any part other than parts p and q.

Depending on the application, a distance function can be defined in many different ways. The most commonly applied distance metrics are the following (Kusiak, 1985).

Minkowski distance metric

$$d_{pq} = \left[\sum_{k=1}^{K} |\mathbf{X}_{pk} - \mathbf{X}_{qk}|^r \right]^{1/r} \tag{2.1}$$

where r is a positive integer. Two special cases of the above metric which are widely used are the

- absolute metric (for $r = 1$)
- euclidean metric (for $r = 2$).

Weighted Minkowski distance metric

$$d_{pq} = \left[\sum_{k=1}^{K} W_k |\mathbf{X}_{pk} - \mathbf{X}_{qk}|^r \right]^{1/r} \tag{2.2}$$

There are two special cases:

- weighted absolute metric (for $r = 1$)
- weighted euclidean metric (for $r = 2$).

Hamming distance metric

$$d_{pq} = \sum_{k=1}^{K} \delta(\mathbf{X}_{pk} \cdot \mathbf{X}_{qk}) \tag{2.3}$$

where

$$\delta(X_{pk}, X_{qk}) = \begin{cases} 1, & \text{if } X_{pk} \neq X_{qk} \\ 0, & \text{otherwise} \end{cases}$$

Example 2.1

A company is using an eight-digit polycode to distinguish part types. Each code digit is assigned a numeric value between 0 and 9. The six part types thus coded are given in Fig. 2.5. Find the Minkowski absolute distance metric between the parts. Determine the Hamming distance metric between parts.

In the Minkowski absolute distance metric given by equation 2.1 ($r = 1$), for example, the distance between parts 1 and 2 is calculated as follows:

$$d_{12} = |X_{11} - X_{21}| + |X_{12} - X_{22}| + |X_{13} - X_{23}| + |X_{14} - X_{24}| + |X_{15} - X_{25}| + |X_{16} - X_{26}| + |X_{17} - X_{27}| + |X_{18} - X_{28}|$$

which gives

$$d_{12} = 1 + 2 + 0 + 1 + 2 + 0 + 1 + 3 = 10$$

Similarly, the distance metrics between all other parts are found and summarized in Fig. 2.6.

Parts \ Digits	1	2	3	4	5	6	7	8
1	3	1	1	6	3	8	0	7
2	4	3	1	5	1	8	1	4
3	4	2	1	5	1	8	0	4
4	5	1	1	6	3	7	0	7
5	4	2	1	5	1	5	1	4
6	3	1	1	6	3	6	2	7

Fig. 2.5 Classification codes of parts.

	1	2	3	4	5	6
1	–	10	8	3	12	4
2	10	–	2	11	4	12
3	8	2	–	9	4	12
4	3	11	9	–	11	5
5	12	4	4	11	–	10
6	4	12	12	5	10	–

Fig. 2.6 Minkowski absolute distance metric between parts.

The Hamming metric is given by equation 2.3. For example, the Hamming metric between parts 1 and 2 is calculated as follows:

$$\mathbf{X}_{11}=3;\quad \mathbf{X}_{21}=4 \quad \Rightarrow \quad \delta(\mathbf{X}_{11},\mathbf{X}_{21})=1$$

$$\mathbf{X}_{12}=1;\quad \mathbf{X}_{22}=3 \quad \Rightarrow \quad \delta(\mathbf{X}_{12},\mathbf{X}_{22})=1$$

$$\mathbf{X}_{13}=1;\quad \mathbf{X}_{23}=1 \quad \Rightarrow \quad \delta(\mathbf{X}_{13},\mathbf{X}_{23})=0$$

$$\mathbf{X}_{14}=6;\quad \mathbf{X}_{24}=5 \quad \Rightarrow \quad \delta(\mathbf{X}_{14},\mathbf{X}_{24})=1$$

$$\mathbf{X}_{15}=3;\quad \mathbf{X}_{25}=1 \quad \Rightarrow \quad \delta(\mathbf{X}_{15},\mathbf{X}_{25})=1$$

$$\mathbf{X}_{16}=8;\quad \mathbf{X}_{26}=8 \quad \Rightarrow \quad \delta(\mathbf{X}_{16},\mathbf{X}_{26})=0$$

$$\mathbf{X}_{17}=0;\quad \mathbf{X}_{27}=1 \quad \Rightarrow \quad \delta(\mathbf{X}_{17},\mathbf{X}_{27})=1$$

$$\mathbf{X}_{18}=7;\quad \mathbf{X}_{28}=4 \quad \Rightarrow \quad \delta(\mathbf{X}_{18},\mathbf{X}_{28})=1$$

$$d_{12}=\sum_{k=1}^{8}\delta(\mathbf{X}_{1k},\mathbf{X}_{2k})=1+1+0+1+1+0+1+1=6$$

Similar calculations between all parts yield the symmetric matrix shown in Fig. 2.7.

2.3 CLUSTER ANALYSIS

The objective of cluster analysis is to assign P parts to f part families while minimizing some measure of distance. The distance measures stored in a two-dimensional array are accessed by a clustering algorithm to group the parts. There are a number of methods which can be used for this purpose. This chapter introduces a hierarchical clustering algorithm, p-median model and a multi-objective clustering algorithm.

Hierarchical clustering algorithm

In this procedure, the parts are first grouped into a few broad families, each of which is then partitioned into smaller part families and so on until the final part families are generated. The parts are clustered at each

	1	2	3	4	5	6
1	–	6	5	2	7	2
2	6	–	2	7	2	7
3	5	2	–	6	2	7
4	2	7	6	–	7	3
5	7	2	2	7	–	7
6	2	7	7	3	7	–

Fig. 2.7 Hamming distance metric between parts.

step by lowering the amount of interaction between each part and a part family mean or median, to develop a tree-like structure called a dendogram. This section illustrates this by considering the 'nearest neighboring approach' (Kusiak, 1983). A number of other procedures are discussed in detail in Chapter 4.

Example 2.2

Using the hierarchical clustering algorithm for the nearest neighboring approach (with Minkowski absolute distance) construct the dendogram (the Minkowski absolute distance is shown in Fig. 2.6).

Iteration 1. Since the objective is to group parts with minimum distance, parts 2 and 3, which have the smallest distance ($d_{23} = 2$) are grouped to form part family {2,3}. The distance between part family {2,3} and the remaining parts is updated as follows:

$$d_{(23)1} = \min\{d_{21}, d_{31}\} = \min\{10,8\} = 8$$

$$d_{(23)4} = \min\{d_{24}, d_{34}\} = \min\{11,9\} = 9$$

$$d_{(23)5} = \min\{d_{25}, d_{35}\} = \min\{4,4\} = 4$$

$$d_{(23)6} = \min\{d_{26}, d_{36}\} = \min\{12,12\} = 12$$

The new distance matrix with revised distances (underlined) is shown in Fig. 2.8 (a). Note that the matrix is symmetric and hence it is sufficient to consider only the upper or lower triangular matrix.

Iteration 2. The smallest distance in the above matrix is between parts 1 and 4. Join them to form the next part family {1,4}. Update the distance between this part family with the other parts and part families as follows:

$$d_{(14)(23)} = \min\{d_{12}, d_{13}, d_{42}, d_{43}\} = \min\{10,8,11,9\} = 8$$

$$d_{(14)5} = \min\{d_{15}, d_{45}\} = \min\{12,11\} = 11$$

$$d_{(14)6} = \min\{d_{16}, d_{46}\} = \min\{4,5\} = 4$$

The revised matrix is shown in Fig. 2.8(b).

Iteration 3. The smallest distance now occurs between part 5 and part family {2,3}. Join them to form a new part family {2,3,5} (part 6 and part family {1,4} could also be selected). The distance matrix is again updated as shown in Fig. 2.8(c).

Iteration 4. The smallest distance is between part 6 and part family {1,4}. Part family {1, 4, 6} is formed and the distance matrix updated as in Fig. 2.8(d). Thus, the distance between the two disjoint part families {1,4,6} and {2,3,5} is 8.

	1	(2,3)	4	5	6
1	–	8	3	12	4
(2,3)		–	9	4	12
4			–	11	5
5				–	10
6					–

	(1,4)	(2,3)	5	6
(1,4)	–	8	11	4
(2,3)		–	4	12
5			–	10
6				–

	(1,4)	(2,3,5)	6
(1,4)	–	8	4
(2,3)		–	10
6			–

	(1,4,6)	(2,3,5)
(1,4,6)	–	8
(2,3,5)		–

Fig. 2.8 Revised Minkowski absolute distance matrix: (a) iteration 1; (b) iteration 2; (c) iteration 3; (d) iteration 4.

Iteration 5. Finally, the two remaining part families are merged with a distance measure of 8. The result of the hierarchical clustering algorithm is shown by a dendogram in Fig. 2.9. The distance scale indicates the distance between sub-clusters at each branching of the tree. The user must decide the distance which best suits the application. For example, if the dendogram is cut at a distance of 6, two part families are formed.

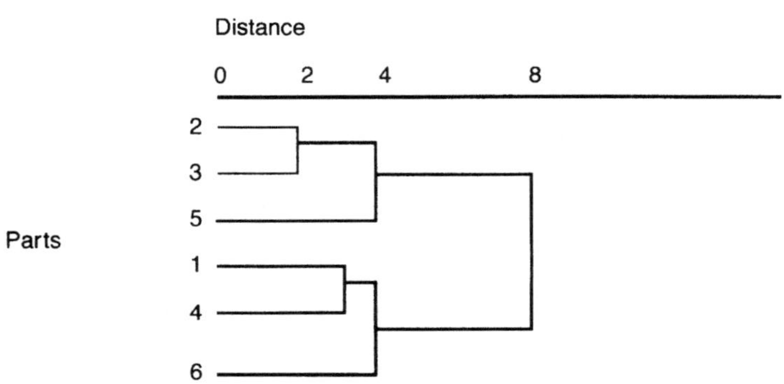

Fig. 2.9 Dendogram showing the distance of parts.

P-median model

This is a mathematical programming approach to the cluster analysis problem. The objective of this model is to identify f part families optimally, such that the distance between parts in each family is minimized with respect to the median of the family. The number of medians f is a given parameter in the model. The model selects f parts as medians and assigns the remaining parts to these medians such that the sum of distances in each part family is minimized. Unlike the hierarchical clustering algorithm, this model allows parts to be transferred from one family to another in order to achieve the optimal solution (Kusiak, 1983). In the following notation P is the number of parts and f the number of part families to be formed. The following relations hold:

$$d_{pq} \geqslant 0, \quad \forall p \neq q, p = 1,2\ldots\ldots,P$$

$$d_{pq} = 0, \quad \forall p = q, p = 1,2\ldots\ldots,P$$

$$\mathbf{X}_{pq} = \begin{cases} 1, & \text{if part } p \text{ belongs to part family } q \\ 0, & \text{otherwise} \end{cases}$$

Minimize

$$\sum_{p=1}^{P} \sum_{q=1}^{P} d_{pq} \mathbf{X}_{pq}$$

subject to:

$$\sum_{q=1}^{P} \mathbf{X}_{pq} = 1, \quad \forall p \tag{2.4}$$

$$\sum_{q=1}^{P} \mathbf{X}_{qq} = f \tag{2.5}$$

$$\mathbf{X}_{pq} \leqslant \mathbf{X}_{qq}, \quad \forall p,q \tag{2.6}$$

$$\mathbf{X}_{pq} = 0/1, \quad \forall p,q \tag{2.7}$$

Constraints 2.4 ensure that each part p is assigned to exactly one part family. The number of part families to be formed is specified by equation 2.5. Constraints 2.6 impose that qth part family is formed only if the corresponding part is a median. If part q is not a median the corresponding $\mathbf{X}_{qq}$ variable takes a value of 0. The last constraints 2.7 ensure integrality. The number of part families to be formed is a parameter in the model.

Example 2.3

By considering the Minkowski absolute distances given in Fig. 2.6 if the p-median model is solved for obtaining two part families, this gives:

$$X_{11} = X_{41} = X_{61} = 1$$

$$X_{33} = X_{23} = X_{53} = 1$$

and all other X_{pq} are zero. Thus, one part family consists of parts {1,4,6} and the other part family consists of parts {3,2,5}. The median parts are 1 and 3, respectively. The objective value is 13.

Multi-objective clustering algorithm

In the effective formation of part families several attributes need to be evaluated according to certain priorities. In the clustering procedure and the p-median model, these n-dimensional attributes were treated as n points. A measure of distance was calculated to represent the dissimilarity for each pair of points. These distances were arranged in a two-dimensional array and used to form part families. In forming GT part families it would be preferable to use a multi-objective approach, where each attribute is evaluated separately by considering some relative importance. This section presents the multi-objective model proposed by Han and Ham (1986) for identifying flexible part families: 'flexible' in the sense that the user has the choice of input digit priority and similarity digit set. Thus, using this model part families can be developed for different applications and the model takes the form

Lex Minimize

$$\sum_{q=1}^{f} \sum_{p=1}^{P} d_{pqk} \cdot X_{pq}, \quad k \in Z$$

subject to:

$$d_{pqk} = 0, \quad \forall k \in z, \; p \text{ for all parts in part family } q \tag{2.8}$$

$$\sum_{q=1}^{f} X_{pq} = 1, \quad \forall p \tag{2.9}$$

$$X_{pq} = 0/1, \quad \forall p, q \tag{2.10}$$

where d_{pqk} is the distance from part p to part family q at digit k,

$$d_{pqk} \geqslant 0, \quad \forall p \neq q, p = 1,2\ldots\ldots,P$$

$$d_{pqk} \geqslant 0, \quad \forall p = q, p = 1,2\ldots\ldots,P$$

Z is the classification code prioritized sequence and z is the set of digits of significant similarity. Constraints 2.8 ensure that all parts in a part family q have the same codes on the significant similarity digits. To ensure a part is assigned to one family, constraints 2.9 are imposed. Constraints 2.10 indicate integer variables.

The objective function of the model lexico-graphically minimizes the distance between digits. This means the distance is minimized according to a sequence in which the user specifies the input prioritized codes. The values in a pair of distance vectors are examined in decreasing order of priority. Lower priorities cannot preempt, or override, a higher priority. The parts are grouped into part families on the significant similarity between digits. All parts in a family have the same codes of significant similarity digit set z. By varying the code digit priorities and significant similarity between digits, part families can be created for diverse applications such as purchasing, tool design, process planning, machine grouping etc. The algorithm is iterative and is similar to the bond energy algorithm (Chapter 3). Initially, within a part family the two most similar parts are found and grouped. Then the part most similar to the first two is found (by lexicographic minimization) and grouped. Next, the part most similar to that one is found and grouped (Gongaware and Ham, 1981). This process is repeated for all part families. However, since the method utilizes goal programming, proper selection of priorities is important to obtain meaningful results.

Example 2.4

From the classification code of parts in Fig. 2.5, use the multi-objective approach to form two part families. The classification code prioritized sequence vector $Z = [4,5,8,1,3,2,7,6]$ and the set of digits of significant similarity is $z = [4,5,8]$. Identify the optimal sequence in which the parts are arranged in each part family.

The rearranged part code based on the prioritized sequence vector Z is shown in Fig. 2.10. The distances between the parts are calculated using the Minkowski absolute distance metric. The calculations used in deriving the part families are presented in Table 2.1. $X_{pq} = 1$ indicates part p is assigned to part family q. The distances are calculated between two consecutive parts in a part family. Two part families $\{1,4,6\}$ and $\{2,3,5\}$ are formed. The multi-objective cluster algorithm is designed to

Parts \ Digits	4	5	8	1	3	2	7	6
1	6	3	7	3	1	1	0	8
2	5	1	4	4	1	3	1	8
3	5	1	4	4	1	2	0	8
4	6	3	7	5	1	1	0	7
5	5	1	4	4	1	2	1	5
6	6	3	7	3	1	1	2	6

Fig. 2.10 Prioritized part code.

Table 2.1 Distance calculations

Prioritized sequence		*Digits*								
q	*p*	*4*	*5*	*8*	*1*	*3*	*2*	*7*	*6*	$\mathbf{X}_{pq}$
	1	0	0	0	0	0	0	0	0	1
	2	1	–							0
	3	1	–							0
1	4	0	0	0	2	0	0	0	1	1
	5	1	–							0
	6	0	0	0	2	0	0	2	1	1
	2	0	0	0	0	0	0	0	0	1
2	3	0	0	0	0	0	1	1	0	1
	5	0	0	0	0	0	0	1	3	1
$\bar{u}_i$		0	0	0	4	0	1	4	5	

optimize lexicographically the sequence of parts within each part family. Here, for purpose of illustration, we will consider all the possible sequences for each part family and determine the optimal sequence. Alternatively, the procedure stated by Gongaware and Ham (1981) could be used. For the part family {1,4,6} the differences arise in the first, seventh and sixth digits. The contribution to the objective function for the six possible sequences with respect to digits 1,7 and 6 is given in Table 2.2(a)–(f). Since lexicographic minimization requires the contribution to $\bar{u}_i$ of the highest ranking digit in the prioritized order be minimized, in this case digit 1, the sequences {1,4,6} and {6,4,1} are eliminated. For the remaining sequences, since the distance is the same we proceed to consider the next significant digit and compute the distances. Based on digit 7 the sequences {1,6,4} and {4,6,1} are not considered. Finally, the two remaining sequences are compared for digit 6. Since both sequences {4,1,6} and {6,1,4} have the same distance, we can select arbitrarily, say {4,1,6}. A similar analysis for the second part family will identify two possible sequences {2,5,3} or {3,5,2}, say {2,5,3}. The optimal arrangement is shown in Table 2.3.

2.4 RELATED DEVELOPMENTS

While a number of companies have used informal techniques to indentify part families, a formal coding and classification system has great potential. Numerous coding systems have been developed all over the world by university researchers, consulting firms and also by corporations for their own use. In a recent survey of 53 respondents in the USA (Wemmerlov and Hyer, 1989), 62% indicated the use of one or more classification schemes in conjuction with GT applications.

Table 2.2 Determining the optimal sequence for part family {1,4,6}

	Digits				Digits		
Sequence	*1*	*7*	*6*	*Sequence*	*1*	*7*	*6*
1	0			1	0	0	
4	2			6	0	2	
6	2			4	2	2	
(a) Distance $\bar{u}_i$	4			(b) Distance $\bar{u}_i$	2	4	
	Digits				Digits		
Sequence	*1*	*7*	*6*	*Sequence*	*1*	*7*	*6*
4	0	0	0	4	0	0	
1	2	0	1	6	2	2	
6	0	2	2	1	0	2	
(c) Distance $\bar{u}_i$	2	2	3	(d) Distance $\bar{u}_i$	2	4	
	Digits				Digits		
Sequence	*1*	*7*	*6*	*Sequence*	*1*	*7*	*6*
6	0	0	0	6	0		
1	0	2	2	4	2		
4	2	0	1	1	2		
(e) Distance $\bar{u}_i$	2	2	3	(f) Distance $\bar{u}_i$	4		

A handful of commercial systems are available in the US market (Hyer and Wemmerlov, 1984). Hyer and Wemmerlov (1985) discussed in detail the different code structures, uses and the guidelines for implementation. In a later article, Hyer and Wemmerlov (1989) presented the results of a survey of 53 GT users, 33 of whom used

Table 2.3 Optimal arrangement

Prioritized sequence		Digits								
q	*p*	*4*	*5*	*8*	*1*	*3*	*2*	*7*	*6*	X_{pq}
	4	0	0	0	0	0	0	0	0	1
1	1	0	0	0	2	0	0	0	1	1
	6	0	0	0	0	0	0	2	2	1
	2	0	0	0	0	0	0	0	0	1
2	5	0	0	0	0	0	1	0	3	1
	3	0	0	0	0	0	0	1	3	1
$\bar{u}_i$		0	0	0	2	0	1	3	9	

coding and classification. For practitioners in the process of selecting and justifying GT software, Wemmerlov (1990) provided information from software vendors, interviews with manufacturers and published sources. Tatikonda and Wemmerlov (1992) reported an empirical study of classification and coding system usage among manufacturers. The investigation, selection, justification, implementation and operation of different systems by six user firms were presented in a case study form. For a list of available classification systems and their sources, see Gallagher and Knight (1986) (p. 133). A number of other methods have also been proposed for part family formation (Knight, 1974; Kusiak, 1985; Dutta *et al.*, 1986; Shiko, 1992). Kasilingam and Lashkari (1990) developed a mathematical model for allocating new parts to existing part families. A number of individual case studies of implementation of coding and classification systems have also been reported (Dunlap and Hirlinger, 1983; Marion, Rubinovich and Ham, 1986; Rajamani, 1993).

2.5 SUMMARY

Part family formation provides a number of benefits in terms of manufacturing, product design, purchasing etc. All parts in a family, depending on the purpose, may require similar treatment, handling, and design features, enabling reducing setup times, improved scheduling, improved process control and standardized process plans. Coding of parts is an important step towards the identification of part families. Although a number of coding schemes are available, research has indicated that a universal system of classification and coding is not practical, although it would be preferable. The complexity of the manufacturing environment is such that a system more tailored to individual needs is essential to provide an accurate database. The development of a coding scheme and the process of coding is expensive and time-consuming. However, GT coding and classification provides the link between design and manufacturing and is an integral and important part of future CAD/CAM activities. Three distance measures which are commonly used as a measure of performance and a few clustering methods for identifying part families (i.e. classifying parts) have been presented. The numerous benefits of GT in a variety of business problems have led at least one user to believe (Hyer and Wemmerlov, 1989): 'the use for GT and its extensive database are limited only by the user's imagination and the problems presented to it'. Here it is important to emphasize that GT is a general philosophy to improve performance throughout the organization; the coding and classification system is only a tool to help implement GT.

PROBLEMS

2.1 What is a part family? What are the benefits of part family formation?
2.2 What is a composite part? Give an example of your own.
2.3 What are the different coding systems and what are their relevance in the context of part family formation?
2.4 What are the main advantages of polycode over monocode?
2.5 An ABC company has established a nine-digit coding scheme to distinguish between various types of parts. The six part types coded are given below. Each code digit is assigned a numeric value between 0 and 9:

part 1: 112171213
part 2: 112175427
part 3: 112174327
part 4: 102173203
part 5: 112175327
part 6: 412174453.

(a) Find the Minkowski absolute distance between the parts.
(b) Using the hierarchical clustering algorithm construct the dendogram for parts.
(c) Identify two part families by defining a suitable threshold value.
(d) Find the Hamming distance metrics between the six part types.

2.6 Consider the Hamming distance metrics between parts in 2.5(*d*).

(a) Using the p-median model identify two part families.
(b) Does the best grouping always correspond to the minimum distance?

2.7 For the classification code of six parts in 2.5, use the multi-objective approach to form two part families. The classification code prioritized sequence vector $Z = [5,9,3,4,2,1,6,7,8]$ and the set of digits of significant similiarity $z = [5,9]$. Identify the optimal sequence in which the parts are arranged in each part family.

REFERENCES

Billo, R.E., Rucker, R. and Shunk, D. L. (1987) Integration of a group technology classification and coding system with an engineering database. *Journal of Manufacturing Systems*, **6**(1), 37–45.

Chevalier, P. W. (1984) Group technology as a CAD/CAM integrator in batch manufacturing. *International Journal of Operations and Production Research*, **3**, 3–12.

Dunlap, G. C. and Hirlinger, C. R. (1983) Well planned coding and classification system offers company wide synergistic benefits. *Industrial Engineering*, November, 78–83.

Dutta, S. P., Lashkari, G., Nadoli, G. and Ravi, T. (1986) A heuristic procedure for determining manufacturing families from design-based grouping for flexible manufacturing systems. *Computers and Industrial Engineering*, **10**(3), 193–201.

Eckert, R. L. (1975) Codes and classification systems. *American Machinist*, December, 88–92.

Fu, K. S. (1980) Recent developments in pattern recognition. *IEEE Transactions on Computers*, **29**(10), 845–54.

Gallagher, C. C. and Knight, W. A. (1986) *Group Technology Production Methods in Manufacture*, Ellis Horwood, Chichester.

Gongaware, T. A. and Ham, I. (1981) Cluster analysis applications for group technology manufacturing systems, *in Proceedings of the IX North American Metal-working Research Conference*. Society of Manufacturing Engineers, Dearborn, MI, pp. 131–6.

Groover, M. P. (1987) *Automation, Production Systems and Computer Integrated Manufacturing*, Kluwer-Nijhoff Publishing, Boston.

Han, C. and Ham, I. (1986) Multiobjective cluster analysis for part family formations. *Journal of Manufacturing Systems*, **5**(4), 223–30.

Hyer, N. L. and Wemmerlov, U. (1984) Group technology and productivity, *Harvard Business Review*, **62**(4), 140–9.

Hyer, N. L. and Wemmerlov, U. (1985), Group technology oriented coding systems: structures, applications and implementation, *Production and Inventory Management*, **26**, 55–78.

Hyer, N. L. and Wemmerlov, U. (1989) Group technology in the US manufacturing industry: a survey of current practices. *International Journal of Production Research*, **27**(8), 1287–304.

Jackson, D. (1978) *Cell System of Production*, Business Books, London.

Kasilingam, R. G. and Lashkari, R. S. (1990) Allocating parts to existing part families in cellular manufacturing systems. *International Journal of Advanced Manufacturing Technology*, **3**, 3–12.

Knight, W. A. (1974) Part family methods for bulk metal forming. *International Journal of Production Research*, **12**(2), 209–31.

Kusiak, A. (1983) Part families selection model for flexible manufacturing systems, in *Proceedings of the Annual Industrial Engineering Conference*, Louisville KY, pp. 575–80.

Kusiak, A. (1985) The part families problem in flexible manufacturing systems. *Annals of Operations Research*, **3**, 279–300.

Marion, D., Rubinovich, J. and Ham, I. (1986) Developing a group technology coding and classification scheme. *Industrial Engineering*, July, 90–7.

Rajamani, D. (1993) Classification and coding of components for implementing a computerized inventory system for a television assembling industry. *International Journal of Production Economics*, **32**, 133–54.

Shiko, G. (1992) A process planning-oriented approach to part family formation in group technology applications. *International Journal of Production Research*, **30**(8), 1739–52.

Tatikonda, M. V. and Wemmerlov, U. (1992) Adoption and implementation of group technology classification and coding systems: insights from seven case studies. *International Journal of Production Research*, **30**(9), 2087–110.

Wemmerlov, U. (1990) Economic justification of group technology software: documentation and analysis of current practices. *Journal of Operations Management*, **9**(4), 500–25.

Wemmerlov, U. and Hyer, N. L. (1989) Cellular manufacturing in the US industry: a survey of users. *International Journal of Production Research*, **27**(9), 1511–30.

CHAPTER THREE

Part–machine group analysis: methods for cell formation

Early applications of group technology used the classification and coding techniques to identify part families. The application areas included design, process planning, sales, purchasing, cost estimation etc. Depending on the application area, the appropriate attributes were selected. A distance measure was then defined followed by the identification of part families using a suitable clustering technique. The emphasis in this and subsequent chapters is on GT application to manufacturing. The simplest application of GT which is common in batch environments is to rely informally on part similarities to gain efficiencies when sequencing parts on machines. The second application is to create formal part families, dedicate machines to these families, but let the machines remain in their original positions (logical layout). The ultimate application is to form manufacturing cells (physical layout). The logical layout is applied when part variety and production volumes are changing frequently such that a physical layout which requires rearrangement of machines is not justified.

Traditionally coding schemes emphasized the capture of part attributes, thus identifying families of parts which were similar in function, shape etc., but gave no help in identifying the set of machines to process them. Burbidge (1989, 1991) proposed production flow analysis (PFA) to find a complete division of all parts into families and also a complete division of all the existing machines into associated groups by analysing information in the process routes for parts. If manufacturing attributes were considered by classification and coding to identify the part families, we believe the division will be similar to that obtained using PFA. However, the main attraction of PFA is its simplicity and it gets results relatively quickly. The appropriateness of PFA against classification and coding in different situations has yet to be fully researched. This chapter discusses some well known algorithms to identify the part families and machine groups which are accomplished manually by PFA. PFA is a systematic procedure for dividing the complete organization. Identification of part families and machine groups discussed in this chapter is one of the steps in PFA.

The identification of part families and machine groups is commonly referred to as cell formation. Numerous approaches have been reported for cell formation. These approaches adopt either a sequential or simultaneous procedure to partition the parts and machines. The sequential procedure determines the part families (or machine groups) first, followed by machine assignment (or part allocation). For example, classification and coding can be used to identify the part families, followed by identification of the machines required to process each part family. The simultaneous procedure determines the part families and machine groups concurrently. PFA and the algorithms presented in this chapter fall into this class of procedure.

3.1 DEFINITION OF THE PROBLEM

The application or adoption of GT starts with identifying part families and machine groups such that each part family is processed within a machine group with minimum interaction with other groups. Cell formation is recognized by researchers as a complex problem, so it often proceeds in stages. There is a need to limit the scope of the problem at each stage because attempts to broaden the problem complicate it and lead to failure (Burbidge, 1993). PFA, which has been successfully applied in at least 36 factories, is based on this philosophy and considers one simple change: the change from process organization to GT. It does not consider changes in plant, product design, processing methods or suboptimizations such as cost minimization at this stage. Some of these are desirable, but they are best left as new projects after GT (Burbidge, 1993). Chapters 3 to 5 consider cell formation as a reorganization of an existing job shop into GT shops using information given about the processing requirements of parts.

The processing requirements of parts on machines can be obtained from the routing cards. This information is commonly represented in a matrix called the part–machine matrix, which is a $P*M$ matrix with 0 or 1 entries. An example of a part–machine matrix is shown in Fig. 3.1. A 1 in column p and row m indicates that part p requires machine m for an operation. The sequence of operations is ignored by this matrix and if a part requires more than one operation on a machine, this cannot be identified in the part–machine matrix (using 0 and 1). Moreover, only the machine types are referred to in the above matrix, not the number of copies available of a given machine type. The basic assumption is that the machine type within the group to which the part is assigned has sufficient capacity to process the parts completely.

		Part (p)				
		1	2	3	4	5
Machine (m)	1	1	0	1	0	0
	2	0	1	1	0	1
	3	1	0	0	1	0
	4	0	0	1	0	1

Fig. 3.1 Part–machine matrix.

		Part					
		1	2	3	4	5	
Machine	1	1	0	1			
	2	0	1	1		1	Exceptional element
	3	1		0	1	0	Void
	4			1	0	1	

Fig. 3.2 Arbitrary partition.

Let us now arbitrarily partition the matrix as shown in Fig. 3.2 to identify two diagonal blocks (cells) which correspond to two part families and machine groups. Parts 1, 2 and machines 1, 2 are in one cell, while parts 3, 5 and machines 3, 4 are in the other cell. With this partition it is observed that parts 1, 3 and 5 visit both cells to complete all operations. This is indicated by the 1s outside the diagonal blocks. These are referred to as 'exceptional' parts, and the machines 1, 2 and 3 which process these parts are identified as 'bottleneck' machines. Also, it is observed that part 2 does not require machine 1, although it is provided in the cell. This is indicated by a 0 inside the diagonal block. Similarly, all the other parts do not require one of the machines assigned to the cells. The 0s inside the diagonal blocks are referred to as 'voids'.

If instead of arbitrarily partitioning the matrix, we interchange the rows and columns, the resulting matrix is shown in Fig. 3.3. In this new partition the numbers of 1s outside and 0s inside are less than the previous partition. Ideally, one would like to partition such that there are no 0s inside the diagonal blocks and no 1s outside the diagonal blocks (Fig. 3.4). This implies that the two cells are independent, i.e. each part family is completely processed within a machine group and each part in a part family is processed by every machine in the corresponding machine group. This example illustrates a case when a perfect decomposition of a system into two subsystems (cells) is obtained. However, in real life the nature of the data set is such that a perfect

		Part				
		1	4	3	5	2
Machine	1	1	0	1		
	3	1	1			
	2			1	1	1
	4			1	1	0

Fig. 3.3 Rearranged partition.

		Part				
		1	4	3	5	2
Machine	1	1	1			
	3	1	1			
	2			1	1	1
	4			1	1	1

Fig. 3.4 Perfect clusters.

decomposition is hardly ever obtained. In this situation (Fig. 3.3), one would like to obtain a near-perfect decomposition considering the following objectives while partitioning the matrix (Miltenburg and Zhang, 1991):

1. to have minimum number of 0s inside the diagonal blocks (voids);
2. to have minimum number of 1s outside the diagonal blocks (exceptional elements).

A void indicates that a machine assigned to a cell is not required for the processing of a part in the cell. When a part passes a machine without being processed on the machine, it contributes to an additional intra-cell handling cost. This leads to large, inefficient cells. An exceptional element is created when a part requires processing on a machine that is not available in the allocated cell of the part. When a part needs to visit a different cell for its processing the inter-cell handling cost increases. This also requires more coordinating effort between cells. Thus, voids and exceptional elements are undesirable. The voids and exceptional elements created are dependent on the number of diagonal blocks and the size of each diagonal block. In general, as the number of diagonal blocks decreases the size of blocks increases. This results in more voids and fewer exceptional elements. If all parts and machines are grouped as one diagonal block (i.e. the cell is large and loose) we have maximum voids and no exceptional elements (Adil, Rajamani and Strong, 1993). For example, the matrix in Fig. 3.1 has 11 voids and no exceptional elements. On the other hand, if the number of diagonal blocks is increased to two, say after rearranging as shown in Fig. 3.3, the voids reduce to two and the exceptional elements increase to one. Thus, as the number of voids are reduced, the number of exceptional elements increases, and vice-versa.

This chapter presents a few matrix manipulation algorithms. These are simple algorithmic procedures for rearranging the rows and columns. Once the matrix is rearranged, the user has to identify the part families and machine groups. Procedures to take care of exceptional parts and bottleneck machines will also be considered. Finally, a number of performance measures will be defined which consider the trade-off between voids and exceptional elements illustrated above. This is followed by a report on the comparative performance of the major algorithms.

3.2 BOND ENERGY ALGORITHM (BEA)

The bond energy algorithm was developed by McCormick, Schweitzer and White (1972) to identify and display natural variable groups or clusters that occur in complex data arrays. They proposed a measure of effectiveness (ME) such that an array that possesses dense clumps of numerically large elements will have a large ME when compared with the same array the rows and columns of which have been permuted so that its numerically large elements are more uniformly distributed throughout the array. The ME of an array A (summed bond energy (BE) over all row and column permutations) is given by

$$\text{ME }(A) = 1/2 \sum_{p=1}^{P} \sum_{m=1}^{M} a_{pm}[a_{p,m+1} + a_{p,m-1} + a_{p+1,m} + a_{p-1,m}] \tag{3.1}$$

with

$$(a_{0,m} = a_{P+1,m} = a_{p,0} = a_{p,M+1} = 0)$$

where

$$a_{pm} = \begin{cases} 1, & \text{if part } p \text{ requires processing on machine } m \\ 0, & \text{otherwise} \end{cases}$$

Maximizing the ME by row and column permutations serves to create strong bond energies, that is,

Maximize

$$\frac{1}{2}\sum_{p=1}^{P}\sum_{m=1}^{M} a_{pm}[a_{p,m+1}+a_{p,m-1}+a_{p+1,m}+a_{p-1,m}]$$

where the maximization is taken over all $P!\,M!$ possible arrays that can be obtained from the input array by row and column permutations. The above equation is also equivalent to

$$\text{ME }(A)=\sum_{m=1}^{M-1}\sum_{p=1}^{P} a_{pm}\cdot a_{p,m+1}+\sum_{p=1}^{P-1}\sum_{m=1}^{M} a_{pm}\cdot a_{p+1,m} \tag{3.2}$$

$$=\text{ME (rows)}+\text{ME (columns)}$$

Since the vertical (horizontal) bonds are unaffected by the interchanging of the columns (rows), the ME decomposes to two parts: one finding the optimal column permutation, the other finding the optimal row permutation. A sequential-selection suboptimal algorithm which exploits the nearest-neighbor feature as suggested by McCormick, Schweitzer and White (1972) is as follows.

Algorithm

Step 1. Select a part column arbitrarily and set $i=1$. Try placing each of the remaining $(P-i)$ part columns in each of the $(i+1)$ possible positions (to the left and right of the i columns already placed) and compute the contribution of each column to the ME:

$$\text{ME (columns)}=\sum_{p=1}^{i}\sum_{m=1}^{M} a_{pm}\cdot a_{p+1.m}$$

Place the column that gives the largest BE in its best position. In case of a tie, select arbitrarily. Increment i by 1 and repeat until $i=P$. When all the columns have been placed, go to step 2.
Step 2. Repeat the procedure for rows, calculating the BE as

$$\text{ME (rows)}=\sum_{m=1}^{i}\sum_{p=1}^{P} a_{pm}\cdot a_{p,m+1}$$

(Note that the row placement is unnecessary if the input array is symmetric, since the final row and column orderings will be identical.)

Example 3.1

Find the measure of effectiveness of the matrix given in Fig. 3.5 using equations 3.1 and 3.2.

		Part 1	Part 2	Part 3	Part 4
Machine	1	1	1	0	0
	2	0	0	1	1
	3	0	0	1	0
	4	1	1	0	0

Fig. 3.5 Part–machine matrix for Example 3.1.

	0	0	0	0	
0	1	1	0	0	0
0	0	0	1	1	0
0	0	0	1	0	0
0	1	1	0	0	0
	0	0	0	0	

ME

1	1	0	0
0	0	2	1
0	0	1	0
1	1	0	0

Fig. 3.6 ME for a_{pm}.

	1	2	3	4
1	1	1	0	0
2	0	0	1	1
3	0	0	1	0
4	1	1	0	0

ME (rows)

0	0	0	0
0	0	1	0
0	0	0	0

ME (columns)

1	0	0
0	0	1
0	0	0
1	0	0

Fig. 3.7 ME for rows and columns.

Using equation 3.1, the ME for a_{pm} is shown in Fig. 3.6, i.e. $\text{ME} = 1/2(1+1+1+1+2+1+1) = 4$. Using equation 3.2 the ME for rows and columns are shown in Fig. 3.7, i.e. ME = ME (rows) + ME (columns) = (1) + (1 + 1 + 1) = 4.

Example 3.2

Consider the matrix of Fig. 3.8 of four parts and four machines.

Step 1. Pick any part column say $p = 1$. Place the other columns at its sides and compute the ME (Table 3.1). Note that the selected columns are underlined and the ME for each placement is shown within brackets. In the case of a tie, select arbitrarily, say in this case (1 3). Again place the remaining columns and compute the ME (Table 3.2). Select (1 3 2) and proceed to place the last column (Table 3.3). Select (1 3 2 4) as the sequence of column placement. The permuted matrix at the end of this step is shown in Fig. 3.9. Proceed to step 2.

		Part			
		1	2	3	4
Machine	1	0	1	0	1
	2	1	0	0	0
	3	1	0	1	0
	4	0	0	0	1

Fig. 3.8 Initial part–machine matrix.

Table 3.1 Computing the ME for $p = 1$

	Column											
	1	2	2	1	1	3	3	1	1	4	4	1
	0	1	1	0	0	0	0	0	0	1	1	0
	1	0	0	1	1	0	0	1	1	0	0	1
	1	0	0	1	1	1	1	1	1	0	0	1
	0	0	0	0	0	0	0	0	0	1	1	0
ME	(0)		(0)		(1)		(1)		(0)		(0)	

Table 3.2 Computing the ME for columns (1 3)

	Column																	
	1	3	2	1	2	3	2	1	3	1	3	4	1	4	3	4	1	3
	0	0	1	0	1	0	1	0	0	0	0	1	0	1	0	1	0	0
	1	0	0	1	0	0	0	1	0	1	0	0	1	0	0	0	1	0
	1	1	0	1	0	1	0	1	1	1	1	0	1	0	1	0	1	1
	0	0	0	0	0	0	0	0	0	0	0	1	0	1	0	1	0	0
ME	(1)			(0)			(1)			(1)			(0)					(1)

Step 2. The above procedure will now be repeated for rows. Select any row, say $m = 1$ (Table 3.4). Select (1 4) and proceed (Table 3.5). Select, say (3 1 4) (Table 3.6). Select either (3 2 1 4) or (2 3 1 4) to obtain the row placement. The final rearranged matrix (one possible solution) is shown in Fig. 3.10.

Limitations of the BEA

The final ordering obtained is independent of the order in which rows (columns) are selected but is dependent on the initial row (column) selected to initiate the process. However, McCormick, Schweitzer and White (1972) reported that the solutions are numerically close when

Table 3.3 Computing the ME for columns (1 3 2)

	Column															
	1	3	2	4	1	3	4	2	1	4	3	2	4	1	3	2
	0	0	1	1	0	0	1	1	0	1	0	1	1	0	0	1
	1	0	0	0	1	0	0	0	1	0	0	0	0	1	0	0
	1	1	0	0	1	1	0	0	1	0	1	0	0	1	1	0
	0	0	0	1	0	0	1	0	0	1	0	0	1	0	0	0
ME	(2)				(2)				(0)				(1)			

		Part 1	3	2	4
Machine	1	0	0	1	1
	2	1	0	0	0
	3	1	1	0	0
	4	0	0	0	1

Fig. 3.9 Column-permuted matrix.

		Part 1	3	2	4
Machine	3	1	1	0	0
	2	1	0	0	0
	1	0	0	1	1
	4	0	0	0	1

Fig. 3.10 Final rearranged matrix.

tried on different starting rows (columns). Since this was developed as a general-purpose clustering algorithm, no discussion on exceptional elements and bottleneck machines has been provided.

3.3 RANK ORDER CLUSTERING (ROC)

This algorithm was developed by King (1980 a, b) for part–machine grouping. It provides a simple, effective and efficient analytical technique which can be easily computerized. In addition, it has fast convergence and a low computation time. Each row (column) in the part–machine matrix is read as a binary word. The procedure

Table 3.4 Computing the ME for $m = 1$

	Row		*Row*		*Row*		*Row*		*Row*		*Row*	
	$\underline{1}$	0 0 1 1	2	1 0 0 0	$\underline{1}$	0 0 1 1	3	1 1 0 0	$\underline{1}$	0 0 1 1	4	0 0 0 1
	2	1 0 0 0	$\underline{1}$	0 0 1 1	3	1 1 0 0	$\underline{1}$	0 0 1 1	4	0 0 0 1	$\underline{1}$	0 0 1 1
ME		(0)		(0)		(0)		(0)		(1)		(1)

Table 3.5 Computing the ME for rows (1 4)

	Row		*Row*		*Row*		*Row*		*Row*		*Row*	
	$\underline{1}$	0 0 1 1	2	1 0 0 0	$\underline{1}$	0 0 1 1	$\underline{1}$	0 0 1 1	3	1 1 0 0	$\underline{1}$	0 0 1 1
	$\underline{4}$	0 0 0 1	$\underline{1}$	0 0 1 1	2	1 0 0 0	$\underline{4}$	0 0 0 1	$\underline{1}$	0 0 1 1	3	1 1 0 0
	2	1 0 0 0	$\underline{4}$	0 0 0 1	$\underline{4}$	0 0 0 1	3	1 1 0 0	$\underline{4}$	0 0 0 1	$\underline{4}$	0 0 0 1
ME		(1)		(1)		(0)		(1)		(1)		(0)

converts these binary words for each row (column) into decimal equivalents. The algorithm successively rearranges the rows (columns) iteratively in order of descending values until there is no change. The algorithm is given below.

ROC algorithm

Step 1. For row $m = 1, 2........, M$, compute the decimal equivalent c_m by reading the entries as binary words:

$$\text{i.e. } c_m = \sum_{p=1}^{P} 2^{P-p}.a_{pm} \quad (a_{pm} = 0 \text{ or } 1)$$

Reorder the rows in decreasing c_m. In the case of a tie, keep the original order.
Step 2. For column $p = 1, 2........., P$, compute the decimal equivalent r_p, by reading the entries as binary words:

$$\text{i.e. } r_p = \sum_{m=1}^{M} 2^{M-m}.a_{pm} \quad (a_{pm} = 0 \text{ or } 1)$$

Reorder the columns in decreasing r_p. In the case of a tie, keep the original order.
Step 3. If the new part–machine matrix is unchanged, then stop, else go to step 1.

Example 3.3

Apply ROC to the part–machine matrix shown in Fig. 3.11.

Step 1. The decimal equivalents of the binary number for rows are given in the right-hand side of the matrix in Fig. 3.12. The rank order of the rows is shown in brackets. On rearranging the rows in order of decreasing rank, the row-permuted matrix is shown in Fig. 3.13.
Step 2. The rank order for columns is also shown in Fig. 3.13. By rearranging the columns in order of decreasing rank, the column-permuted matrix is shown in Fig. 3.14.

Step 1. Rearrange the rows based on the rank order as shown in Fig. 3.14 to obtain the matrix shown in Fig. 3.15.
Step 2. Further rearrangement of columns does not occur based on the ranking shown in Fig. 3.15.
Step 3. On performing steps 1 and 2, the matrix remains unchanged, therefore stop.

From the block diagonal matrix shown in Fig. 3.15, there are few possible ways one can identify the part families and machine groups. Two such possibilities are shown in Figs. 3.16 and 3.17, respectively. The

Table 3.6 Computing the ME for rows (3 1 4)

	Row					*Row*					*Row*					*Row*				
	3	1	1	0	0	3	1	1	0	0	3	1	1	0	0	2	1	0	0	0
	1	0	0	1	1	1	0	0	1	1	2	1	0	0	0	3	1	1	0	0
	4	0	0	0	1	2	1	0	0	0	1	0	0	1	1	1	0	0	1	1
	2	1	0	0	0	4	0	0	0	1	4	0	0	0	1	4	0	0	0	1
ME		(1)					(0)					(2)					(2)			

	Part 1	2	3	4	5	6	7	8
Machine 1	1	1			1			1
2			1	1		1	1	
3	1	1	1		1	1		1
4				1			1	1
5			1	1			1	
6	1	1				1		

	2^7	2^6	2^5	2^4	2^3	2^2	2^1	2^0	
	Part 1	2	3	4	5	6	7	8	
Machine 1	1	1			1			1	2^5
2			1	1		1	1		2^4
3	1	1	1		1	1		1	2^3
4				1			1	1	2^2
5			1	1			1		2^1
6	1	1				1			2^0

Fig. 3.11 Initial part–machine matrix for Example 3.3.

Binary weight

	2^7	2^6	2^5	2^4	2^3	2^2	2^1	2^0	Decimal equivalent	Rank
	Part 1	2	3	4	5	6	7	8		
Machine 1	1	1			1			1	201	(2)
2			1	1		1	1		54	(4)
3	1	1	1		1	1		1	237	(1)
4				1			1	1	19	(6)
5			1	1			1		50	(5)
6	1	1				1			196	(3)

Fig. 3.12 Step 1, computing the decimal equivalents.

two-cell arrangement (Fig. 3.16) leads to the minimum number of exceptional elements and voids and hence is selected.

Limitations of ROC

1. The reading of entries as binary words presents computational difficulties. Since the largest integer representation in most computers is $2^{48} - 1$ or less, the maximum number of rows and columns is restricted to 47.

Machine \ Part	1	2	3	4	5	6	7	8	Binary weight
3	1	1	1		1	1		1	2^5
1	1	1			1			1	2^4
6	1	1				1			2^3
2			1	1		1	1		2^2
5			1	1			1		2^1
4				1			1	1	2^0
Decimal equivalent	56	56	38	7	48	44	7	49	
Rank	(1)	(2)	(6)	(7)	(4)	(5)	(8)	(3)	

Fig. 3.13 Step 1, row-permuted matrix.

	2^7	2^6	2^5	2^4	2^3	2^2	2^1	2^0		
Machine \ Part	1	2	8	5	6	3	4	7	Decimal equivalent	Rank
3	1	1	1	1	1	1			252	(1)
1	1	1	1	1					240	(2)
6	1	1			1				200	(3)
2					1	1	1	1	15	(5)
5						1	1	1	7	(6)
4			1				1	1	37	(4)

Fig. 3.14 Step 2, column-permuted matrix.

2. The results are dependent on the initial matrix, so the final solution is not necessarily the best solution. This also makes the treatment of exceptional elements arbitrary.
3. It has a tendency to collect 1s in the top left-hand corner, while the rest of the matrix is disorganized.
4. Even in well structured matrices it is not certain ROC will identify the block diagonal structure.
5. The identification of bottleneck machines and exceptional parts is arbitrary and is crucial to the identification of subsequent groupings.

3.4 RANK ORDER CLUSTERING 2 (ROC 2)

ROC 2 was developed by King and Nakornchai (1982) to overcome the computational limitations imposed by ROC. This algorithm begins by identifying in the right-most column all rows that have an entry of 1.

Machine \ Part	1	2	8	5	6	3	4	7	Binary weight
3	1	1	1	1	1	1			2^5
1	1	1	1	1					2^4
6	1	1			1				2^3
4			1				1	1	2^2
2					1	1	1	1	2^1
5						1	1	1	2^0
Decimal equivalent	56	56	52	48	42	35	7	7	
Rank	(1)	(2)	(3)	(4)	(5)	(6)	(7)	(8)	

Fig. 3.15 Step 1 (iteration 2).

Machine \ Part	1	2	8	5	6	3	4	7
3	1	1	1	1	1	1		
1	1	1	1	1	0			
6	1	1	0	1	1			
4			1			0	1	1
2					1	1	1	1
5						1	1	1

Number of exceptional elements = 3
Number of voids = 3
Total = 6

Fig. 3.16 Part family and machine groups (arrangement 1).

Machine \ Part	1	2	8	5	6	3	4	7
3	1	1	1	1	1	1		
1	1	1	1	1				
6	1	1		1	1			
4			1			0	1	1
2					1	1	1	1
5						1	1	1

Number of exceptional elements = 7
Number of voids = 1
Total = 8

Fig. 3.17 Part family and machine groups (arrangement 2).

These rows are moved to the top of the column, keeping the relative order among rows. This procedure is then applied to the rows by beginning at the last row. The use of binary words is eliminated in this procedure, but the idea of rank ordering still remains with the other limitations. The procedure is also implemented in an interactive program with various facilities to rearrange the data in the manner required. Thus, even for very complicated matrices, various trial assignments of exceptional elements and transfer of parts of the same type can be made and the results can be quickly determined. If the outcome is not as expected a return to the previous stage can be carried out quickly and another trial conducted. The algorithm is given below.

ROC 2 algorithm

Step 1. Row arrangement. From $p = P$ (the last column) to 1, locate the rows with an entry of 1; move the rows with entries to the head of the row list, maintaining the previous order of entries.
Step 2. Column arrangement. From $m = M$ (the last row) to 1, locate the columns with an entry of 1; move the columns with entries to the head of the column list, maintaining the previous order of entries.
Step 3. Repeat steps 1 and 2 until no change occurs or inspection is required.

Example 3.4

Consider the matrix of eight parts and six machines in Fig. 3.11.

Step 1. Row arrangement. Select the last column $p = 8$. The initial order of rows is 1,2,3,4,5,6. Underscore the rows which contain a 1 in column 8, to obtain 1,2,3,4,5,6. Move 1,3 and 4 to the head of the list followed by 2,5 and 6 in the same order as read from left to right. Do this for all columns (Table 3.7). The row-permuted matrix is shown in Fig. 3.18. For convenience again renumber these rows as 1 to 6 starting from the top. Thus the new row 1 actually corresponds to the old row 3, and so on.
Step 2. Column arrangement. The above process is now repeated by selecting the last row $m = 6$ (old row 4). The sequence of shifting is shown in Table 3.8 (the old row numbers are shown in brackets). The column-permuted matrix is shown in Fig. 3.19.

Step 1. Row arrangement. Rearrange the rows to observe if there is a change in order (Table 3.9). The revised row permuted matrix is shown in Fig. 3.20.
Step 2. Column arrangement. The column arrangement does not change.
Step 3. No further improvement is possible, hence stop.

The final matrix obtained is the same as that obtained using ROC.

Treatment for bottleneck machines

The procedure proposed by King and Nakornchai (1982) for bottleneck machines is as follows:

Table 3.7 Step 1, row arrangement

Columns	*Rows*					
8	1	2	3	4	5	6
7	1	3	4	2	5	6
6	4	2	5	1	3	6
5	2	3	6	4	5	1
4	3	1	2	6	4	5
3	2	4	5	3	1	6
2	2	5	3	4	1	6
1	3	1	6	2	5	4
ρ	3	1	6	2	5	4

Machine \ Part	1	2	3	4	5	6	7	8	New row numbers
3	1	1	1		1	1		1	1
1	1	1			1			1	2
6	1	1				1			3
2			1	1		1	1		4
5			1	1			1		5
4				1			1	1	6

Fig. 3.18 Row-arranged matrix.

Step 1. Simply ignore the bottleneck machines (rows). This has the slight effect of decreasing the problem size.
Step 2. Apply the ROC 2 algorithm to the remainder problem.
Step 3. Depending on the number of copies of bottleneck machines available, various block diagonal combinations are possible. Based on judgement (can consider providing a copy to a cell which processes maximum parts), assign copies to each cell.
Step 4. Apply ROC 2 to this extended problem.

Step 3 makes it possible to experiment with alternate merging as well as taking account of various practical constraints in determining a feasible solution.

Treatment for exceptional elements

The formal procedure for dealing with exceptional elements is as follows (King and Nakornchai, 1982):

Step 1. Use ROC (ROC 2) to generate the diagonal structure.

Table 3.8 Step 2, column arrangement

Rows	*Columns*							
6(4)	1	2	3	4	5	6	7	8
5(5)	4	7	8	1	2	3	5	6
4(2)	4	7	3	8	1	2	5	6
3(6)	4	7	3	6	8	1	2	5
2(1)	6	1	2	4	7	3	8	5
1(3)	1	2	8	5	6	4	7	3
Column arrangement	1	2	8	5	6	3	4	7

Machine \ Part	1	2	8	5	6	3	4	7	New row numbers
3	1	1	1	1	1	1			1
1	1	1	1	1					2
6	1	1			1				3
2					1	1	1	1	4
5						1	1	1	5
4			1				1	1	6
New column numbers	1	2	3	4	5	6	7	8	

Fig. 3.19 Row- and column-arranged matrix.

Step 2. Identify the exceptional elements.
Step 3. Temporarily ignore these by replacing the 1s by a 0 and continue the ROC (ROC 2) algorithm.
Step 4. Reinstate the exceptional elements in the final matrix designating them by an asterisk instead of 1.

3.5 MODIFIED RANK ORDER CLUSTERING (MODROC)

The fact that ROC has a tendency to collect all the 1s in the top left-hand corner was identified by Chandrasekaran and Rajagopalan (1986a). By removing this block of columns from the matrix and performing ROC again, MODROC collects another set of 1s in the top left-hand corner. This process is continued until no elements are left in the matrix. This process will identify mutually exclusive part families but may contain overlapping machines. A hierarchical clustering method is applied based on a measure of association among pairs of machine groups. Clustering is terminated when the groups are non-intersecting or when a single group is formed. In the latter case the number of groups is

Table 3.9 Step 1 (pass 2), row arrangement

Columns	*Rows*					
8(7)	1(3)	2(1)	3(6)	4(2)	5(5)	6(4)
7(4)	4	5	6	1	2	3
6(3)	4	5	6	1	2	3
5(6)	4	5	1	6	2	3
4(5)	4	1	3	5	6	2
3(8)	1	2	4	3	5	6
2(2)	1	2	6	4	3	5
1(1)	1	2	3	6	4	5
Row order	1(3)	2(1)	3(6)	6(4)	4(2)	5(5)

Part

Machine	1	2	8	5	6	3	4	7	New row numbers
3	1	1	1	1	1	1			1
1	1	1	1	1					2
6	1	1			1				3
4			1				1	1	4
2					1	1	1	1	5
5						1	1	1	6
New column numbers	1	2	3	4	5	6	7	8	

Fig. 3.20 Row arrangement (revised).

determined on a suitable decision criterion and the bottleneck machines are identified at the appropriate hierarchical level in the clustering process. The algorithm is presented below.

MODROC algorithm

Step 1. Apply ROC to the matrix and perform both row and column iterations.

Step 2. Identify the largest block of 1s in the top left-hand corner of the matrix as follows (Chandrasekaran and Rajagopalan, 1986a). Initiate a search procedure from a_{11} through a_{22} to a_{pp} until a zero is encountered. Then p is decremented to $(p-1)$ and the search progresses along the row until a 0 is encountered along $a_{p-1,m}$. Then m is decremented to $(m-1)$ and the block is identified with $a_{p-1,m-1}$ as its last elements. If the search along the row $a_{p-1,m}$ does not change the block from the square shape, it progresses along the columns in a similar manner. If both searches are unsuccessful, the obvious choice is the square block with $a_{p-1,m-1}$ as its last elements.

Step 3. Store the part family and machine group.
Step 4. Slice away the columns corresponding to the block.
Step 5. Go to step 1 and iterate until all columns are grouped.
Step 6. Generate the lower triangular matrix s_{ij}, where s_{ij} is the measure of association between groups C_i and C_j and is defined as the ratio of the number of common elements to the number of elements in the smaller group, i.e. $s_{ij} = n(C_i \cap C_j) / n(\text{Min}\,[C_i, C_j])$.
Step 7. Locate the highest s_{ij} and join group i and j; the corresponding part families; print the result.
Step 8. Update s_{ij} and check Max (s_{ij}). If equal to zero, go to step 10.
Step 9. Go to step 7 and iterate until the number of groups is one.
Step 10. Stop.

The application of steps 1–5 is similar to ROC, except, ROC has to be performed a number of times with progressively smaller matrices. Steps 6–9 correspond to the hierarchical clustering algorithm. The application of this procedure is dealt with in Chapter 4.

3.6 DIRECT CLUSTERING ALGORITHM (DCA)

Chan and Milner (1982) proposed the DCA, which rearranges the rows with the left-most positive cells (i.e. 1s) to the top and the columns with the top-most positive cells to the left of the matrix. Wemmerlov (1984) provided a correction to the original algorithm to get consistent results. The revised algorithm is given below.

Algorithm

Step 1. Count the number of 1s in each column and row. Arrange the columns in decreasing order by placing them in a sequence as identified by starting from the last element (right-most), while moving towards the first (left). Similarly, arrange rows in increasing order in a sequence as identified by starting from the last element (bottom-most), while moving towards the first (top). (Note that this rearrangement of the initial matrix has been proposed to ensure that the final solution would always be the same.)
Step 2. Start with the first column of the matrix. Pull all the rows with 1s to the top to form a block. If, on considering subsequent columns, the rows with 1s are already in the block, do nothing. If there are rows with 1s not in the block, let these rows form a block and move this block to the bottom of the previous block. Once assigned to a block, a row will not be moved; thus, it may not be necessary to go through all the columns.
Step 3. If the previous matrix and current matrix are the same, stop, else go to step 4.

Step 4. Start with the first row of the matrix and pull all the columns to the left (similar to step 2).
Step 5. If the previous matrix and current matrix are the same, stop, else go to step 2.

Example 3.5

Apply the DCA to the part–machine matrix in Fig. 3.21.

Step 1. The number of 1s in each column and row are shown in Fig. 3.21. On rearranging, the sequence of columns and rows are [3,6,4,1,5,2] from left to right and [5, 4, 3, 2, 1] from top to bottom. The rearranged matrix is shown in Fig. 3.22.
Step 2. Against the first column (3) move the block of rows [5,4,2] to the top left-hand corner. Since the second column corresponding to 6 has a 1 in rows 5 and 4, which already exist in the first block, the second block consists of rows [3,1] obtained by considering the column corresponding to part 4. Since all rows are assigned to a block, the blocks are placed as [5,4,2][3,1] from top to bottom. The remaining columns need not be scanned. The rows thus arranged are shown in Fig. 3.23.
Step 3. Since the current matrix differs from the previous matrix, proceed to step 4.
Step 4. Instead of moving rows, in this step we will move the columns. The first block will thus be against row(5) and is [3,6]. The subsequent blocks are [5], [4,1] and [2]. The matrix thus rearranged is shown in Fig. 3.24.

Limitation of the DCA

This procedure, again, may not necessarily always produce diagonal solutions, even if one exists. For example, if the DCA is performed on data in Fig. 3.11, it leads to an unacceptable solution.

	Part						Number of 1s
	1	2	3	4	5	6	
Machine 1	1	1		1			3
2			1		1		2
3	1			1			2
4			1			1	2
5			1			1	2
Number of 1s	2	1	3	2	1	2	

Fig. 3.21 Initial part–machine matrix for Example 3.5.

		Part						Number of 1s
		3	6	4	1	5	2	
Machine	5	1	1					2
	4	1	1					2
	3			1	1			2
	2	1				1		2
	1			1	1		1	3
Number of 1s		3	2	2	2	1	1	

Fig. 3.22 Step 1, rearranged matrix.

		Part					
		3	6	4	1	5	2
Machine	5	1	1				
	4	1	1				
	2	1				1	
	3			1	1		
	1			1	1		1

Fig. 3.23 Step 2, rearranged matrix.

		Part					
		3	6	5	4	1	2
Machine	5	1	1				
	4	1	1				
	2	1		1			
	3				1	1	
	1				1	1	1

Fig. 3.24 Step 4, rearranged matrix.

3.7 CLUSTER IDENTIFICATION ALGORITHM (CIA)

Iri (1968) suggested one of the simplest methods to identify perfect block diagonals if they exist, using a masking technique. This may be described as follows. Starting from any row, mask all the columns which have an entry of 1 in this row, then proceed to mask all the rows which have an entry of 1 in these columns. Repeat the process until the numbers of rows and columns stop increasing. These rows and columns constitute a block. If perfect block diagonals do not exist, the entire matrix is masked as one group. Kusiak and Chow (1987) proposed the CIA as an implementation of this procedure. It is not designed to decompose a matrix to a near-block diagonal form, but simply to identify disconnected blocks if there are any. The algorithm is given below.

Algorithm

Step 1. Select any row m of the matrix and draw a horizontal line h_m through it.
Step 2. For each entry of 1 on the intersection with the horizontal line h_m, draw a vertical line v_p.
Step 3. For each entry 1 crossed by vertical line v_p, draw a horizontal line h_m.
Step 4. Repeat steps 2 and 3 until there are no single-crossed entries 1 left. All double-crossed entries 1 form the corresponding machine group and part family.
Step 5. Transform the original matrix by removing the rows and columns corresponding to machine groups and part family identified in step 4. Rows and columns dropped do not appear in subsequent iterations.
Step 6. If no elements are left in the matrix, stop, else consider the transformed matrix and go to step 1.

Example 3.6

Consider the matrix in Fig. 3.21 and apply the CIA.

Step 1. Select row 1 arbitrarily and draw a horizontal line h_1.
Step 2. Draw vertical lines v_1, v_2 and v_4 intersecting h_1.
Step 3. Draw horizontal line h_3 crossing the 1 entries by v_1 and v_4. This result is shown in Fig. 3.25.
Step 4. Since there are no single-crossed entries, the double-crossed entries 1 form the first machine group (1,3) and part family (1,2,4).
Step 5. The transformed matrix obtained by deleting the machine rows and part columns corresponding to the first cell is shown in Fig. 3.26.
Step 6. Since there are elements still remaining in the matrix, repeat steps 1–4. The resultant matrix identifying the second machine group and part family is also shown in Fig. 3.26. Since after this iteration, no elements are left, stop. The final clustering result is shown in Fig. 3.27.

Limitations of the CIA

As mentioned above, due to the nature of the data if the matrix is not mutually separable, the CIA will mask the complete matrix. Although computationally attractive, it has limited use.

Machine \ Part	1	2	3	4	5	6	
1	1	1		1			h_1
2			1		1		
3	1			1			h_3
4			1			1	
5			1			1	
	v_1	v_2		v_4			

Fig. 3.25 Results of steps 1–3 in the CIA.

Machine \ Part	3	5	6		
2	1	1		h_2	
4	1		1		h_4
5	1		1	h_5	
	v_3	v_5	v_6		

Fig. 3.26 Transformed matrix.

Machine \ Part	1	2	4	3	5	6
1	1	1	1			
3	1		1			
2				1	1	
4				1		1
5				1		1

Fig. 3.27 Final clustering result.

3.8 MODIFIED CIA

In the CIA procedure proposed by Kusiak and Chow (1987), each element of the matrix is scanned twice. Boctor (1991) proposed a new method where each element of the matrix is scanned only once. The algorithm as proposed is given below.

Algorithm

Step 1. Select any machine *m* and the parts visiting it and assign it to the first cell.
Step 2. Consider any other machine. It will be assigned based on one of the following rules:

(a) If none of the parts processed by this machine is already assigned to any cell already created, create a new cell and assign the machines and parts to the new cell.
(b) If some parts are already assigned to one, and only one, other cell, assign the machine and parts to the same cell.
(c) If the parts processed by this machine are assigned to more than one cell, group all these parts and machines together to create a new cell, and add the machines and parts to this cell.

Step 3. Repeat step 2 until all the machines are assigned.

Example 3.7

Illustrate the application of the modified CIA on the matrix in Fig. 3.21.

Step 1. Select machine 1 and parts 1, 2 and 4 and assign it to cell 1.
Step 2. Select, say, machine 2; since parts 3 and 5 are not assigned according to step 2(a), assign them to a new cell 2.
Step 3. Since all the machines are not yet assigned go to step 2.
Step 2. Select, say, machine 3; since parts 1 and 4 processed on this machine are already assigned to cell 1 according to step 2(b), assign machine 3 and the parts to cell 1.
Step 3. Repeat step 2.
Step 2. Select machine 4; since part 3 is already assigned to cell 2 and part 6 is not assigned to any cell according to step 2(b), assign machine 4 and parts 3 and 6 to cell 2.
Step 3. Repeat step 2.
Step 2. Select the last machine 5; since parts 3 and 6 are already assigned to cell 2, assign machine 5 also to cell 2.
Step 3. Since all machines are assigned, stop.

Thus, machines 1,3 and parts 1,2,4 are assigned to cell 1, while machines 2,4,5 and parts 3,5,6 are assigned to cell 2. The partition thus obtained is the same as that shown in Fig. 3.27. This algorithm also carries over the limitations of CIA.

3.9 PERFORMANCE MEASURES

To compare the quality of solutions obtained by different algorithms on an absolute scale, there is a need to develop performance measures or criteria. This section discusses five measures which have been proposed in the literature. The first three measures require the identification of part families and machine groups, while the other two measures only require the rearranged matrix. A brief discussion on these measures follows the notation.

Notation

$|A|$ number of elements in set A
M number of machines
P number of parts
C number of cells (diagonal blocks)
d number of 1s in the diagonal blocks
e number of exceptional elements in the solution
M_c set of machines in cell c
P_c set of parts in cell c
m index for machine
p index for part
o number of 1s in the matrix $\{a_{pm}\}$
c index for cell
v number of voids in the solution

Given a final part–machine matrix with C identifiable cells.

$$o = \sum_{p=1}^{P} \sum_{m=1}^{M} a_{pm}$$

$$d = \sum_{c=1}^{C} \sum_{p \in P_c} \sum_{m \in M_c} a_{pm}$$

$$v = \sum_{c=1}^{C} |M_c||P_c| - d$$

$$e = o - d$$

Grouping efficiency η

Proposed by Chandrasekaran and Rajagopalan (1986b), this was one of the first measures to evaluate the final result obtained by different algorithms. The 'goodness' of solution depends on the utilization of machines within a cell and inter-cell movement. Grouping efficiency was therefore proposed as a weighted average of the two efficiencies η_1 and η_2:

$$\eta = w\eta_1 + (1 - w)\eta_2 \tag{3.3}$$

where

$$\eta_1 = \frac{o - e}{o - e + v}$$

$$\eta_2 = \frac{MP - o - v}{MP - o - v + e}$$

$$\eta = (w)\frac{o - e}{o - e + v} + (1 - w)\frac{MP - o - v}{MP - o - v + e}$$

A value of 0.5 is recommended for w. η_1 is defined as the ratio of the number of 1s in the diagonal blocks to the total number of elements in the blocks (both 0s and 1s). Similarly, η_2 is the ratio of the number of 0s in the off-diagonal blocks to the total number of elements in the off-diagonal blocks (both 0s and 1s). The weighting factor allows the designer to alter the emphasis between utilization and inter-cell movement.

Limitations of η

1. If $w = 0.5$, the effect of inter-cell movement (exceptional elements) is never reflected in the efficiency values for large and sparse matrices.
2. The range of values for the grouping efficiency normally varies from 75 to 100%. Thus, even a very bad solution with large number of exceptional elements will give values around 75%, giving an unrealistic definition of the zero point.
3. When there is no inter-cell movement, $\eta = 2 \neq 0$.

Grouping efficacy τ

The grouping efficacy was proposed by Kumar and Chandrasekaran (1990) to overcome the low discriminating power of the grouping efficiency between well-structured and ill-structured matrices. It has a more meaningful 0–1 range. Unlike the grouping efficiency, the grouping efficacy is not affected by the size of the matrix:

$$\tau = \frac{1-\psi}{1+\phi} = \frac{o-e}{o+v} \tag{3.4}$$

where $\psi = e/o$ and $\phi = v/o$.

Thus, zero efficacy is the point when all the 1s are outside the diagonal blocks. An efficacy of unity implies a perfect grouping with no exceptional elements and voids. However, the influence of exceptions and voids is not symmetric. Consider the following analysis:

$$\tau = (1-\psi)/(1+\phi)$$

$$\mathrm{d}\tau = -1/(1+\phi)d\psi - (1-\psi)/(1+\phi)^2 d\phi$$

Since the coefficients of $d\psi$ and $d\phi$ are both negative an increase in exceptional elements (ψ) or voids (ϕ) will reduce the value of the grouping efficacy. Also, the coefficient of $d\psi$ is always higher than $d\phi$. Thus, the change in exceptional elements has a greater influence than the change in the number of voids in the diagonal blocks. Finally, the voids in the diagonal blocks become less and less significant at lower efficacies.

Grouping measure η_g

This measure was proposed by Miltenburg and Zhang (1991) and follows from the work of Chandrasekaran and Rajagopalan (1986b). It is also a direct measure of the effectiveness of an algorithm to obtain a final grouped matrix. The value of η_g is high if the utilization of machines is high (fewer voids) and few parts require processing on machines in more than one cell (fewer exceptional elements).

The grouping measure η_g is given by

$$\eta_g = \eta_u - \eta_m, \quad -1 \leqslant \eta_g \leqslant 1 \tag{3.5}$$

where

$$\eta_u = d/(d+v), \quad 0 \leqslant \eta_u \leqslant 1$$

and

$$\eta_m = 1 - (d/o), \quad 0 \leqslant \eta_m \leqslant 1$$

η_u is a measure of usage of parts in the part–machine cell. Large values occur when each part in a given cell uses most of the machines in the group; η_m is a measure of part movement between two cells. Small values of η_m occur when few parts require processing by machines outside their cell. Thus, to maximize η_g, large values of η_u and small values of η_m are preferred. According to this definition, if there is no inter-cell movement of parts the value of $\eta_m = 0$. This has been considered as a primary measure by Miltenburg and Zhang (1992) when comparing the performance of a number of algorithms. They also provided two other measures which could be used to enrich comparisons: the clustering measure and the bond energy measure.

Example 3.8

For the two possible partitions shown in Figs 3.16 and 3.17, compute the three measures discussed above

From Table 3.10 it can be observed that the discriminating power of grouping efficiency is low; it also gives a much higher value in comparison to the other two measures.

Clustering measure η_c

The objective of the algorithms proposed in this chapter is to bring all the non-zero elements around the diagonal; thus, another way to measure the effectiveness is to examine how closely the 1s cluster around the diagonal, i.e.

$$\eta_c = \frac{\left\{\sum_{\text{all } a_{pm}=1} \sqrt{(\delta_h^2(a_{pm}) + \delta_v^2(a_{pm}))}\right\}}{\sum_{p=1}^{P} \sum_{m=1}^{M} a_{pm}} \tag{3.6}$$

where $\delta_h(a_{pm})$ is the horizontal distance between a non-zero element a_{pm} and the diagonal:

$$\delta_h = m - \frac{p(M-1)}{(P-1)} - \frac{(P-M)}{(P-1)}$$

and $\delta_v(a_{pm})$ is the vertical distance between a non-zero element a_{pm} and the diagonal:

$$\delta_v = p - \frac{m(P-1)}{(M-1)} + \frac{(P-M)}{(M-1)}$$

The denominator in equation 3.6 normalizes the measure, since the numerator will increase as the number of machines and parts increase. The horizontal and vertical distances to be computed are illustrated in Fig. 3.28.

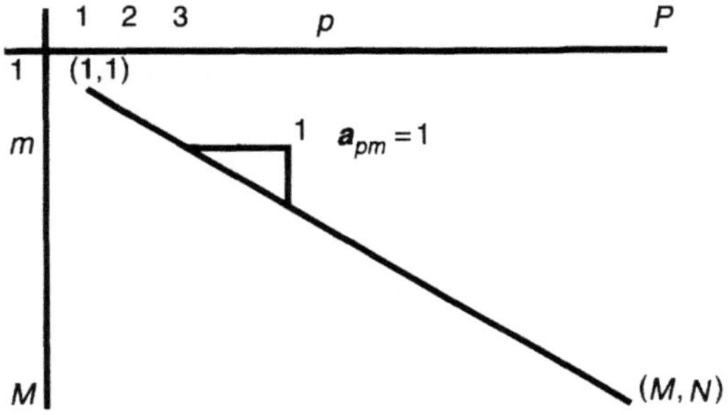

Fig. 3.28 Distance calculation.

Bond energy measure η_{BE}

This measure is based on the premise of the bond energy algorithm, which aims at bringing all the non-zero elements as close together as possible. This is defined as

$$\eta_{BE} = \frac{\left(\sum_{m-1}^{M-1} \sum_{p-1}^{P} a_{pm} \cdot a_{p,m+1} + \sum_{p-1}^{P-1} \sum_{m-1}^{M} a_{pm} . a_{p+1,m}\right)}{\sum_{P=1}^{P} \sum_{m=1}^{M} a_{pm}} \tag{3.7}$$

The above measure is a normalized expression of the measure of effectiveness proposed by McCormick, Schweitzer and White (1972) Normalizing permits comparison of different problems. Large values of η_{BE} are preferred, although at times it is difficult to interpret a high value of the bond energy (Miltenburg and Zhang, 1991). However, as the objective is to compare the quality of solutions it enriches the analysis.

Example 3.9

Compute the clustering and bond energy measures for Figs. 3.8 and 3.10.

The values obtained in Table 3.11 indicate that the clustering measure is low for a block diagonalized matrix (Fig. 3.10) in comparison with the initial matrix (Fig. 3.8). A sample calculation to illustrate the computation of the clustering measure for Fig. 3.8 is given in Table 3.12.

$$\left(M = 4;\ P = 4;\ \delta_h = m - p;\ \delta_v = p - m;\ \delta_h^2 + \delta_v^2 = 2(p - m)^2;\ \sum_{p=1}^{4} \sum_{m=1}^{4} a_{pm} = 6\right)$$

$$\eta_c = \frac{\left\{\sum_{\text{all } a_{pm}=1} \sqrt{(\delta_h^2(a_{pm}) + \delta_v^2(a_{pm}))}\right\}}{\sum_{p=1}^{4} \sum_{m=1}^{4} a_{pm}} = \frac{\sqrt{2} + \sqrt{18} + \sqrt{2} + \sqrt{8}}{6} = 1.6499$$

Table 3.10 Calculation of the three performance measures

Performance measure	*Fig. 3.16*	*Fig. 3.17*
	$M = 6; P = 8; o = 24; e = 3; v = 3; d = 21$	$M = 6; P = 8; o = 24; e = 7; v = 1; d = 17$
Grouping efficiency (equation 3.3)	$0.4375 + 0.4375 = 0.875$	$0.472 + 0.383 = 0.855$
Grouping efficacy (equation 3.4)	0.778	0.68
Grouping measure (equation 3.5)	$0.875 - 0.125 = 0.75$	$0.944 - 0.292 = 0.652$

3.10 COMPARISON OF MATRIX MANIPULATION ALGORITHMS

The computational complexity of the BEA and ROC is $O(PM^2 + P^2M)$, while that of the CIA is $O(2PM)$. However, what is more important is the ability of the algorithms to arrive at a good block diagonal form irrespective of the nature of data set, whether the data are perfectly groupable or not. Chu and Tsai (1990) compared the BEA, ROC and the DCA on 11 data sets from the literature. They compared the performance based on the following four measures:

- total bond energy (equation 3.1);
- percentage of exceptional elements (number of exceptional elements/total number of 1 entries);
- machine utilization (only η_1 of equation 3.3);
- grouping efficiency (equation 3.3).

They summarized their results as follows:

1. No matter which measure of performance of data set is tested, the BEA is the best under evaluation.
2. If a data set is well structured, all three methods can almost completely cluster parts into part families.
3. If exceptional elements exist in the data set, it is much more efficient and effective to use the BEA because the method does not require an additional procedure to arrive at better results.
4. If bottleneck machines exist, none of the three methods can produce acceptable clusters without additional processing.
5. Finally, the BEA not only performs better than ROC and the DCA, it can compete with other methods in the literature, especially if a company wants to reduce the percentage of exceptional elements and increase the 'clumpiness' of the clustering.

3.11 RELATED DEVELOPMENTS

Only a few well known matrix manipulation procedures have been discussed in this chapter. A number of other procedures have also been developed for the same purpose. Khator and Irani (1987) introduced the 'occupancy value' method for progressively developing a block diagonal matrix starting from the north-west corner of the matrix. Ng (1991) showed that the bond energy formulation is equivalent to solving two rectilinear traveling salesman problems. He also established a new worst-case bound for this problem. Ng (1993) proposed several policies to improve the grouping efficiency and efficacy. Kusiak (1991) proposed

Table 3.11 Calculation of the clustering and bond energy measures

Performance measure	*Fig. 3.6*	*Fig. 3.8*
Clustering measure (equation 3.6)	1.6499	0.707
Bond energy measure (equation 3.7)	$1/6 = 0.167$	$4/6 = 0.667$

Table 3.12 Sample calculation of the clustering measure

$p, m \forall a_{pm} = 1$	$\delta_h^2 + \delta_v^2 (\forall a_{pm} = 1) = 2(p - m)^2$
1,2	$2(1-2)^2 = 2$
1,4	18
2,1	2
3,1	8
3,3	0
4,4	0

three algorithms based on different branching schemes for solving the structured and unstructured matrix with restrictions on the number of machines in each cell. Each algorithm uses the CIA concept. Boe and Cheng (1991) proposed a 'close neighbor' algorithm. These are just a few methods; the list is by no means comprehensive.

3.12 SUMMARY

The primary objective of cell formation is to group parts and machines such that all the parts in a family are processed within a machine group with minimum interaction with other groups. If the problem is one of reorganizing existing facilities, information on machine requirements for each part can be obtained from the routing cards. This information is often summarized in the form of a part–machine matrix. The problem now is to identify the part families and machine groups by rearranging the matrix in a block diagonal form, with a minimum number of parts traveling between cells. This is an NP-hard problem. In this chapter, a number of efficient algorithms were presented for manipulating the matrix, to obtain a near-block diagonal form. These procedures (excluding the BEA) require the identification of bottleneck machines and exceptional parts before obtaining the near-block diagonal form, and subsequently identify part families and machine groups.

A number of performance measures were presented which could be used to decide on the best partition. This process requires manual/subjective human intervention. In fact it is difficult to represent and visualize clusters for matrices with large numbers of rows and columns. Moreover, these procedures are unable to consider multiple copies of the same machine type; also, they do not consider other manufacturing aspects such as part sequence, processing times, production volumes, capacity of machines etc. However, these procedures are 'quick and dirty' in the sense that they are easy to construct and to obtain data for. This generates a first-cut solution, and the exceptional elements and each group can be individually considered for a more detailed analysis that integrates other manufacturing aspects. The main feature which makes these algorithms attractive is the fact that they simultaneously group parts and machines. The next chapter introduces a few traditional clustering techniques.

PROBLEMS

3.1 What is cell formation? What are the objectives of cell formation?

3.2 What is an ideal cell? Discuss the implications of exceptional elements and voids in the context of an ideal cell.

3.3 How do the permutations of rows and columns serve to create 'strong bond energies'?

3.4 Consider the part–machine matrix given in Fig. 3.29. Apply the bond energy algorithm to obtain the final rearranged matrix. Compute the mean effectiveness of the initial and final matrices.

3.5 The following data are provided by a local wood manufacturer. The company is interested in decreasing material handling by changing from a process layout to a GT layout. It proposes to install a conveyor for moving parts within a cell. However, it wishes to restrict the movement of parts between cells. Identify the appropriate performance measure to compare different solutions (i.e. different groupings visible in the rearranged matrix). The rearrangement of the part–machine matrix (Fig. 3.30) can be performed using either ROC or the DCA.

	*p*1	*p*2	*p*3	*p*4
*m*1	1	1		
*m*2			1	
*m*3	1			1
*m*4		1	1	1

Fig. 3.29 Part–machine matrix for Q3.4.

	*p*1	*p*2	*p*3	*p*4	*p*5	*p*6	*p*7	*p*8
*m*1		1			1			1
*m*2				1			1	
*m*3					1			1
*m*4		1					1	
*m*5	1					1		
*m*6	1		1			1		1

Fig. 3.30 Part–machine matrix for wood manufacturer example.

	*p*1	*p*2	*p*3	*p*4	*p*5	*p*6	*p*7	*p*8
*m*1			1	1	1	1		1
*m*2	1		1		1		1	1
*m*3		1		1				
*m*4			1			1		
*m*5	1				1		1	
*m*6				1				
*m*7		1		1	1			
*m*8			1					1

Fig. 3.31 Part–machine matrix for Q3.7.

3.6 Can the CIA be applied to the matrix in 3.5? Why or why not?

3.7 Consider the part–machine matrix given in Fig. 3.31. Apply ROC2 to this matrix and identify the part families and machine groups. Compute the following measures for the initial and rearranged final matrix: grouping efficiency, grouping efficacy, grouping measure, clustering measure and bond energy measure (use $w = 0.5$). Compare the values of grouping efficiency, grouping efficacy and grouping measure. Which in your opinion is a more discriminating indicator and why? Discuss the main difference between grouping efficiency and grouping measure.

REFERENCES

Adil, G.K., Rajamani, D. and Strong, D. (1993) AAA: an assignment allocation algorithm for cell formation. Univ. Manitoba, Canada. Working paper.

Boctor, F.F. (1991) A linear formulation of the machine-part cell formation problem. *International Journal of Production Research*, **29**(2), 343–56.

Boe, W.J. and Cheng, C.H. (1991) A close neighbor algorithm for designing cellular manufacturing systems. *International Journal of Production Research*, **29**(10), 2097–116.

Burbidge, J.L. (1989) *Production Flow Analysis for Planning Group Technology*, Oxford Science Publications, Clarendon Press, Oxford.

Burbidge, J.L. (1991) Production flow analysis for planning group technology. *Journal of Operations Management*, **10**(1), 5–27.

Burbidge, J.L. (1993) Comments on clustering methods for finding GT groups and families. *Journal of Manufacturing Systems*, **12**(5), 428–9.

Chan, H.M. and Milner, D.A. (1982) Direct clustering algorithm for group formation in cellular manufacture. *Journal of Manufacturing Systems*, **1**(1), 64–76.

Chandrasekaran, M.P. and Rajagopalan, R. (1986a) MODROC: an extension of rank order clustering of group technology. *International Journal of Production Research*, **24**(5), 1221–33.

Chandrasekaran, M.P. and Rajagopalan, R. (1986b) An ideal seed non-hierarchical clustering algorithm for cellular manufacturing. *International Journal of Production Research*, **24**(2), 451–64.

Chu, C.H. and Tsai, M. (1990) A comparison of three array based clustering techniques for manufacturing cell formation. *International Journal of Production Research*, **28**(8), 1417–33.

Iri, M. (1968) On the synthesis of loop and cutset matrices and the related problems. *SAAG Memoirs*, **4**(A-XIII), 376.

Khator, S.K. and Irani, S.A. (1987) Cell formation in group technology: a new approach. *Computers and Industrial Engineering*, **12**(2), 131–42.

King, J.R. (1980a) Machine–component grouping in production flow analysis: an approach using rank order clustering algorithm. *International Journal of Production Research*, **18**(2), 213–32.

King, J.R. (1980b) Machine–component group formation in group technology. *OMEGA*, **8**(2), 193–9.

King, J.R. and Nakornchai, V. (1982) Machine–component group formation in group technology: review and extension. *International Journal of Production Research*, **20**(2), 117–33.

Kumar, C.S. and Chandrasekaran, M.P. (1990) Grouping efficacy: a quantitative criterion for goodness of block diagonal forms of binary matrices in group technology. *International Journal of Production Research*, **28**(2), 233–43.

Kusiak, A. (1991) Branching algorithms for solving the group technology problem. *Journal of Manufacturing Systems*, **10**(4), 332–43.

Kusiak, A. and Chow, W.S. (1987) Efficient solving of group technology problem. *Journal of Manufacturing Systems*, **6**(2), 117–24.

McCormick, W.T., Schweitzer, P.J. and White, T.W. (1972) Problem decomposition and data reorganization by a clustering technique. *Operations Research*, **20**(5), 993–1009.

Miltenburg, J. and Zhang, W. (1991) A comparative evaluation of nine well known algorithms for solving the cell formation problem in group technology. *Journal of Operations Management*, **10**(1), 44–72.

Ng, S.M. (1991) Bond energy, rectilinear distance and a worst case bound for the group technology problem. *Journal of the Operational Research Society*, **42**(7), 571–8.

Ng, S.M. (1993) Worst-case analysis of an algorithm for cellular manufacturing. *European Journal of Operational Research*, **69**(3), 384–98.

Wemmerlov, U. (1984) Comments on direct clustering algorithm for group formation in cellular manufacturing. *Journal of Manufacturing Systems*, **3**(1), vii–ix.

CHAPTER FOUR

Similarity coefficient-based clustering: methods for cell formation

'Clustering' is a generic name for a variety of mathematical methods which can be used to find out which objects in a set are similar. Several thousand articles have been published on cluster analysis. It has been applied in many areas such as data recognition, medicine, biology, task selection etc. Most of these applications used certain methods of hierarchical cluster analysis. This is also true in the context of part/machine grouping. The methods of hierarchical cluster analysis follow a prescribed set of steps (Romesburg, 1984), the main ones being the following.

- Collect a data matrix the columns of which stand for objects (parts or machines) to be cluster-analysed and the rows of which are the attributes that describe the objects (machines or parts). Optionally the data matrix can be standardized. Since the input matrix is binary, the data matrix never needs to be standardized in this chapter.
- Using the data matrix, compute the values of a resemblance matrix coefficient to measure the similarity (dissimilarity) among all pairs of objects (parts or machines).
- Use a clustering method to process the values of the resemblance coefficient, which results in a tree, or dendogram, that shows the hierarchy of similarities among all pairs of objects (parts or machines). The clusters can be read from the tree.

Although, the basic steps are constant, there is wide latitude in the definition of the resemblance matrix and choice of clustering method. A resemblance coefficient can be a similarity or a dissimilarity coefficient. The larger the value of similarity coefficient, the more similar the two parts/machines are; the smaller the value of a dissimilarity coefficient, the more similar the parts/machines. A few of the clustering methods which will be discussed are single linkage clustering, average linkage clustering, complete linkage clustering and linear cell clustering.

Methods to decide on the number of groups more objectively considering costs and procedures for assigning copies of machines will also be discussed. This chapter adopts a sequential approach to cell formation. First, machine groups are identified, followed by the part families to be processed in these groups.

4.1 SINGLE LINKAGE CLUSTERING (SLC)

McAuley (1972) was the first to apply single linkage clustering to cluster machines. The data matrix we will cluster-analyse is the part–machine matrix. A similarity coefficient is first defined between two machines in terms of the number of parts that visit each machine. Since the matrix has binary attributes, four types of matches are possible. A two-by-two table showing the number of 1–1, 1–0, 0–1, 0–0 matches between two machines is shown in Fig. 4.1

		Machine n	
		1	0
Machine m	1	a	b
	0	c	d

Fig. 4.1 2 × 2 machines table.

where a is the number of parts visiting both machines, b is the number of parts visiting machine m and not n, c is the number of parts visiting machine n and not m, and d is the number of parts not visiting either machine.

Let S_{mn} denote the similarity between machines m and n. To compute S_{mn}, compare two machine rows m and n, computing the values of a,b,c, and d. A number of coefficients have been proposed which differ in the function of these values. The Jaccards coefficient is most often used in this context. This is written as

$$S_{mn} = a/(a+b+c), \quad 0.0 \leqslant S_{mn} \leqslant 1.0 \tag{4.1}$$

The numerator indicates the number of parts processed on both machines m and n, and the denominator is the sum of the number of parts processed on both machines m and n and the number of parts processed on either machine m or n. The Jaccard coefficient indicates maximum similarity when the two machines process the same part types, in which case $b = c = 0$ and $S_{mn} = 1.0$. It indicates maximum dissimilarity when the two machines do not process the same part types, in which case $a = 0$ and $S_{mn} = 0.0$.

Once the similarity coefficients have been determined for machine pairs, SLC evaluates the similarity between two machine groups as follows: the pair of machines (or a machine and a machine group, or two machine groups) with the highest similarity are grouped together. This process continues until the desired number of machine groups has been obtained or all machines have been combined in one group. The detailed algorithm is given below.

SLC algorithm

Step 1. Compute the similarity coefficient S_{mn} for all machine pairs (using equation 4.1). Assume each machine is in a separate machine group.
Step 2. Find the maximum value in the resemblance matrix. Join the two machine groups (two machines, a machine and a machine group or two machine groups). At each stage, machine group m' and n' are merged into a new group, say t. This new group consists of all the machines in both the groups. Add the new group t and update the resemblance matrix by computing the similarity between the new machine group t and some other machine group v as

$$S_{tv} = \underset{\substack{m \in t \\ n \in v}}{\text{Max}} \{S_{mn}\} \tag{4.2}$$

Remove machine groups m' and n' from the resemblance matrix. At each iteration the resemblance matrix gets eaten away by 1. (For example, consider two machine groups (2,4,5) and (1,3). To determine the group to which machine 6 should be assigned, compute $S_{6(245)} = \max(S_{62}, S_{64}, S_{65})$ and $S_{6(13)} = \max(S_{61}, S_{63})$. Machine 6 will be identified with the group it is most similar to. If S_{61} is a maximum, the new group is (1,3,6). The new similarity matrix is determined between the two groups (1,3,6) and (2,4,5) while 6 and (1,3) are removed from the matrix.)
Step 3. When the resemblance matrix consists of one machine group, stop; otherwise go to step 2.

Example 4.1

Apply SLC to the initial part–machine matrix given in Fig. 3.11.

Step 1. The Jaccards similarity coefficient between machine pairs is computed and is shown in Fig. 4.2(a). For example, the similarity between machines 1 and 4 is $S_{1,4} = 1/(1+5) = 0.167 \cong 0.17$.
Step 2. The maximum value corresponds to the (2,5) machine pair. Join these two machines into a new group and update the resemblance matrix as shown in Fig. 4.2(b) where the similarity between the new group (2,5) and the remaining groups is computed as follows:

$$S_{1(2,5)} = \text{Max}\{S_{12}, S_{15}\} = 0$$

$$S_{(2,5)3} = \text{Max}\{S_{23}, S_{53}\} = 0.25$$

(a)

	1	2	3	4	5	6
1	0	0	0.67	0.17	0	0.40
2		0	0.25	0.40	0.75	0.17
3			0	0.125	0.125	0.5
4				0	0.5	0
5					0	0
6						0

(b)

	1	2.5	3	4	6
1	0	0	0.67	0.17	0.4
2,5		0	0.25	0.50	0.17
3			0	0.125	0.5
4				0	0
6					0

(c)

	1,3	2,5	4	6
1,3	0	0.25	0.17	0.5
2,5		0	0.5	0.17
4			0	0
6				0

(d)

	1,3,6	2,5	4
1,3,6	0	0.25	0.17
2,5		0	0.5
4			0

(e)

	1,3,6	2,5,4
1,3,6	0	0.25
2,5,4		0

Fig. 4.2 (a) Jaccards similarity coefficient computed from Fig. 3.11; (b) updated resemblance matrix for the (2,5) machine pair; (c) updated resemblance matrix joining machines 1 and 3; (d) revised matrix joining machine groups (1,3) and 6; (e) revised matrix joining (2,5) and 4.

$$S_{(2,5)4} = \text{Max}\ \{S_{24}, S_{54}\} = 0.50$$

$$S_{(2,5)6} = \text{Max}\ \{S_{26}, S_{56}\} = 0.17$$

At this step, join the machines 1 and 3 at a similarity level of 0.67, and proceed to update the resemblance matrix (Fig. 4.2(c)). Similarly, now join the machine groups (1,3) and 6 at a similarity level of 0.5 (machine pair (2,5) and 4 could also have been selected). Note that this is the maximum value between any two machines in the group. There could be other machines which have a very low level of similarity yet are combined into one group. This is the major disadvantage of SLC. The revised matrix is shown in Fig. 4.2(d). At this stage, join (2,5) and 4 at level 0.5. Revise the matrix again (Fig. 4.2(e)). Finally join the final two groups at a level of 0.25. The dendogram for this is shown in Fig. 4.3.

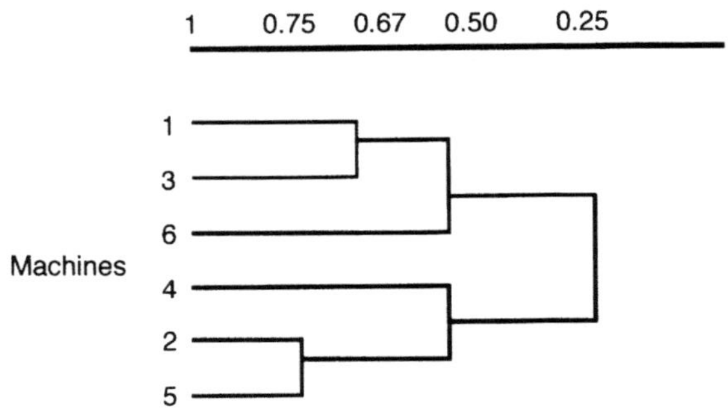

Fig. 4.3 Dendogram for machines using SLC.

4.2 COMPLETE LINKAGE CLUSTERING (CLC)

The complete linkage method combines two clusters at minimum similarity level, rather than at maximum similarity level as in SLC. The algorithm, however, remains the same except that equation 4.2 is replaced by

$$S_{tv} = \underset{\substack{m \in t \\ n \in v}}{\text{Min}} \{S_{mn}\} \tag{4.3}$$

Example 4.2

Apply CLC to the initial part–machine matrix given in Fig. 3.11.

Step 1. As in Example 4.1.
Step 2. The maximum value corresponds to the (2,5) machine pair. Join these two machines into a new group and update the resemblance matrix as shown in Fig. 4.4.(a) where the similarity between the new group (2,5) and remaining groups are computed as follows:

$$S_{1(2,5)} = \text{Min}\ \{S_{12}, S_{15}\} = 0$$

$$S_{(2,5)3} = \text{Min}\ \{S_{23}, S_{53}\} = 0.125$$

$$S_{(2,5)4} = \text{Min}\ \{S_{24}, S_{54}\} = 0.40$$

$$S_{(2,5)6} = \text{Min}\ \{S_{26}, S_{56}\} = 0$$

At this step, join machines 1 and 3 at a similarity level of 0.67 and proceed to update the resemblance matrix (Fig. 4.4(b)). Similarly, now join the machine groups (1,3) and 6 at a similarity level of 0.4 (machine pair (2,5) and 4 could also have been selected). The revised matrix is shown in Fig. 4.4(c). At this stage join (2,5) and 4 at level 0.4. Revise the matrix again (Fig. 4.4(d)). Finally, join the final two groups at a level of 0. The dendogram for this is shown in Fig. 4.5.

(a)

	1	2,5	3	4	6
1	0	0	0.67	0.17	0.4
2,5		0	0.125	0.40	0
3			0	0.125	0.5
4				0	0
6					0

(b)

	1,3	2,5	4	6
1,3	0	0	0.125	0.4
2,5		0	0.4	0
4			0	0
6				0

(c)

	1,3,6	2,5	4
1,3,6	0	0	0
2,5		0	0.4
4			0

(d)

	1,3,6	2,5,4
1,3,6	0	0
2,5,4		0

Fig. 4.4 (a) CLC resemblance matrix computed from Fig. 3.11; (b) updated CLC matrix joining machines 1 and 3; (c) revised CLC matrix joining machine groups (1,3) and 6; (d) revised CLC matrix joining (2,5) and 4.

4.3 AVERAGE LINKAGE CLUSTERING (ALC)

SLC and CLC are clustering based on extreme values. Instead, it may be of interest to cluster by considering the average of all links within a cluster. The initial entries in the S_{mn} matrix consist of similarities associated with all pairwise combinations formed by taking each machine separately. Before any mergers, each cluster consists of one machine. When clusters t and v are merged, the sum of pairwise similarity between the two clusters is

$$AS_{tv} = \frac{\sum_{m \in t} \sum_{n \in v} S_{mn}}{N_t \times N_v} \tag{4.4}$$

where the double summation is the sum of pairwise similarity between all machines of the two groups, and N_t, N_v are the number of machines in groups t and v, respectively. For example, suppose group t consists of machines 1 and 2, group v consists of machines 3, 4 and 5. Then $N_t = 2, N_v = 3$ and

$$AS_{(12)(345)} = (S_{13} + S_{14} + S_{15} + S_{23} + S_{24} + S_{25})/2*3$$

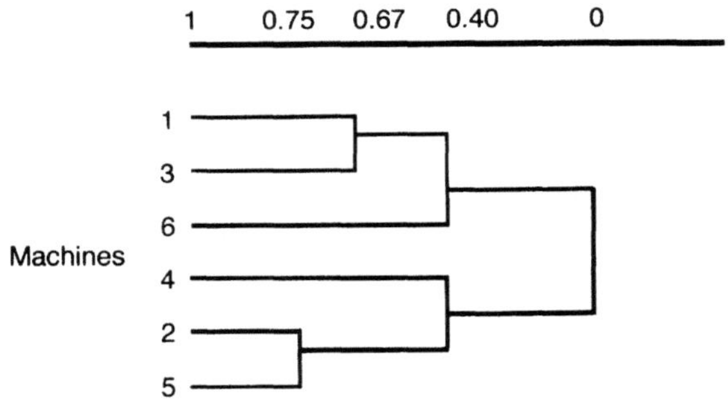

Fig. 4.5 Dendogram for machines using CLC.

Example 4.3

Apply ALC to the initial part–machine matrix given in Fig. 3.11.

Step 1. As in Example 4.1.
Step 2. The maximum value corresponds to the (2,5) machine pair. Join these two machines into a new group and update the resemblance matrix as shown in Fig. 4.6(a). The similarities between the new group (2,5) and remaining groups are computed as follows:

$$S_{1(2,5)} = (S_{12}, S_{15})/(N_1 {*} N_{25}) = (0 + 0)/(1{*}2) = 0$$

$$S_{(2,5)3} = (0.25 + 0.125)/(1{*}2) = 0.19$$

$$S_{(2,5)4} = (0.4 + 0.5)/(2{*}1) = 0.45$$

$$S_{(2,5)6} = (0.17 + 0)/(2{*}1) = 0.084$$

At this stage join machines 1 and 3 at a similarity level of 0.67 and proceed to update the resemblance matrix (Fig. 4.6(b)). Similarly, now join the machine groups (1,3) and 6 at a similarity level of 0.45 (machine pair (2,5) and 4 could also have been selected). Note that this is the maximum value between any two machines in the group. The revised matrix is shown in Fig. 4.6(c). At this stage, join (2,5) and 4 at level 0.45. Revise the matrix again (Fig. 4.6(d)). Finally, join the final two groups at a level of 0.093. The dendogram for this is shown in Fig. 4.7.

In general, the trees produced by these clustering methods will merge machines at different values of the resemblance coefficient, even in cases where they merge the machines in the same order. The above examples

(a)

	1	2.5	3	4	6
1	0	0	0.67	0.17	0.4
2,5		0	0.19	0.45	0.084
3			0	0.125	0.5
4				0	0
6					0

(b)

	1,3	2,5	4	6
1,3	0	0.094	0.146	0.45
2,5		0	0.45	0.084
4			0	0
6				0

(c)

	1,3,6	2,5	4
1,3,6	0	0.09	0.097
2,5		0	0.45
4			0

(d)

	1,3,6	2,5,4
1,3,6	0	0.093
2,5,4		0

Fig. 4.6 (a) ALC resemblance matrix computed from Fig. 3.11; (b) updated ALC resemblance matrix joining machines 1 and 3; (c) revised ALC matrix joining (1,3) and 6; (d) revised ALC matrix joining (2,5) and 4.

illustrate this. SLC produces compacted trees; CLC extended trees; and ALC trees are intermediate between these extremes.

Limitations of SLC, CLC and ALC

1. As a result of SLC, two groups are merged together merely because two machines (one in each group) have high similarity. If this process continues with lone machines that have not yet been clustered, it results in chaining. SLC is the most likely to cause chaining. Since CLC is the antithesis of SLC, it is least likely to cause chaining. ALC produces results between these extremes. When the chaining occurs while machines are being clustered, it is referred to as the 'machine chaining' problem. The following sections address methods to overcome some of these problems.
2. Although the algorithms provide different sets of groups, they do not denote which of these is the best way to group machines. Also, the part families need to be determined.

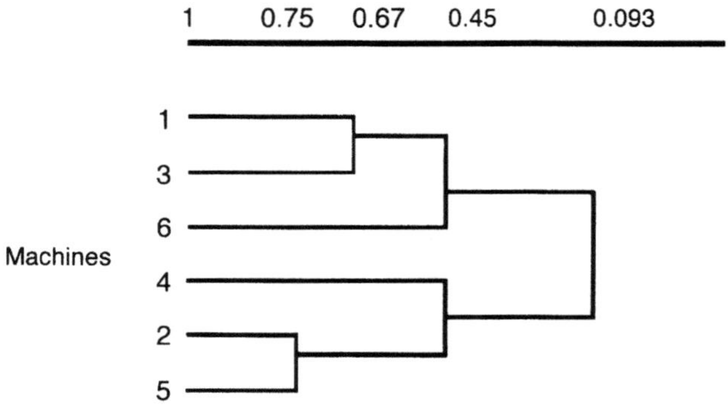

Fig. 4.7 Dendogram for machines using ALC.

3. No insight is provided for the treatment of bottleneck machines.
4. The Jaccards similarity does not give importance to the parts that do not need processing by the machine pairs.

The following sections address methods to overcome a few of these problems.

4.4 LINEAR CELL CLUSTERING (LCC)

The linear cell clustering algorithm was proposed by Wei and Kern (1989). It clusters machines based on the use of a commonality score which defines the similarity between two machines. The commonality score not only recognizes the parts which require the two machines for processing, but also the parts which do not require both the machines. The procedure is flexible and can be adapted to consider constraints pertaining to cell size and number. The worst-case computational complexity of the algorithm is $O(M^2/2 \log (M^2/2) + (M^2/2))$ and is not linear as the name suggests (Chow, 1991; Wei and Kern, 1991). The commonality score and the algorithm are presented below.

Commonality Score

$$C_{mn} = \sum_{p=1}^{P} \delta(a_{pm}, a_{pn}) \tag{4.5}$$

where

$$\delta(a_{pm}, a_{pm}) = \begin{cases} (P-1), & \text{if } a_{pm} = a_{pn} = 1 \\ 1, & \text{if } a_{pm} = a_{pn} = 0 \\ 0, & \text{if } a_{pm} \neq a_{pn} \end{cases}$$

Algorithm

Step 1. Compute the commonality score matrix C_{mn} for all machine pairs.
Step 2. Select the highest score, say corresponding to (m,n). Depending on the state of the two machines, perform one of the following four steps:

(a) If neither machine m nor n is assigned to any group, create a new group with these two machines.
(b) If machine m is already assigned to a group, but not n, then add machine n to the group to which machine m is assigned.
(c) If machines m and n are already assigned to the same group, ignore this score.
(d) If machines m and n are assigned to two different groups, this signifies the two groups may be joined in later processing. Reserve this score for future use.

Step 3. Repeat step 2 until all M machines are assigned to a group.
Step 4. At this stage, the maximum number of clusters that would fit the given matrix is generated. This solution is optimal if the input matrix is perfectly decomposable with no bottleneck parts. However, if the desired number of clusters has not been identified, combine one or more clusters by referring to the scores stored in step 2(d).
Step 5. Select the highest scores among those stored in step 2(d). If the highest score refers to two machines (m,n), combine the two machine groups with machines m and n. If the resultant group is too large, or does not satisfy any of the established constraints, do not join the two machine groups. Instead, select the next-highest score identified in step 2(d). Continue this process until all constraints on the number of groups, group size or cost have been met.

Example 4.4

Apply LCC to the initial part–machine matrix given in Fig. 3.11.

Step 1. The commonality scores between machine pairs are computed and shown in Fig. 4.8. For example, $S_{1,3} = 28 + 2 = 30$ (four 1 and two 0 matches).
Steps 2 and 3. The machines are joined by arranging the scores in descending order based on any of the four steps 2(a)–(d). For this example the different steps carried out at different score levels are shown in brackets, along with the score: 30(2.1), 25(2.1), 23(2.2), 18(2.2), 17(2.3), 17(2.3), and the remaining scores are 14, 9, 9, 7, 7, 2, 2, 1, 0 (2.4).
Step 4. Based on the clustering performed in steps 2 and 3, machines 1,3 and 6 are combined in one group and machines 2, 5 and 4 form the other group. The solution for this example, although not different from that obtained by other algorithms, illustrates the computation involved. Since this grouping leads to few exceptional elements, the solution need not be optimal. Since the desired number of groupings is two, we stop.

	1	2	3	4	5	6
1	0	0	30	9	1	17
2		0	14	17	25	9
3			0	7	7	23
4				0	18	2
5					0	2
6						0

Fig. 4.8 Commonality scores computed from Fig. 3.11.

However, if the number of groups were more than the desired number, the scores which were marked by step 2(d) will be considered, whereby two machine groups will be combined.

4.5 MACHINE CHAINING PROBLEM

The similarity coefficient and commonality score-based methods bring similar machines together. However, in some cases a bottleneck machine may have more common operations with machines in a group other than its assigned group. This improper machine assignment can be reduced by reassigning the bottleneck machine to its proper group. To do so, the number of inter-cellular moves between each bottleneck machine and the machine groups interacting with it are determined. Then, the bottleneck machine is assigned to the group the parts of which have the largest number of operations on the machine. This simple procedure, as suggested by Seifoddini (1989b), totally eliminates the improper machine assignment problem. The primary reason this problem arises is because all the similarity-based methods consider each step of the machine grouping independently. For example, two groups of machines are grouped strictly based on the similarity between two machines m and n without considering their interaction with all other machines. Although the ALC algorithm reduces this problem by considering the average interaction of all the machines in a group, it does not completely eliminate the problem. To address this problem Chow (1992) introduced the machine unit concept and proposed grouping machines using the LCC algorithm. In the machine unit concept, every preceding step is an input to the next step of the solution. For instance, if machines m and n are grouped to form a cell c, then in the next iteration the cell c is transformed into a single unit of machine. To illustrate this concept, assume for two machines m and n the a_{pm} and a_{pn} vector for six parts is given as:

$$m = (1, 0, 1, 0, 1, 1)$$

$$n = (1, 1, 0, 0, 1, 0)$$

The new machine unit c is achieved according to the following rule:

$$a_{p,(mn)} = \begin{cases} 1 & \text{if } a_{pm} = 1 \text{ or } a_{pn} = 1 \\ 0 & \text{otherwise} \end{cases} \tag{4.6}$$

The machine unit c is $=(1,1,1,0,1,1)$.
The following algorithm considering the machine unit concept was proposed by Chow (1992).

Algorithm

Step 1. Compute the commonality scores (equation 4.5) of the part–machine matrix.
Step 2. Group machines m and n with the highest commonality score.
Step 3. Transform machines in step 2 into a new machine unit c as defined in equation 4.6. Replace machines m and n with machine c in the part–machine matrix.
Step 4. If the desired number of groups is formed or the number of machine units is one in the revised part–machine matrix, then stop, otherwise proceed to step 1.

It is important to point out that the grouping process proceeds similarly to the LCC algorithm, i.e. ungrouped machines are given first priority for grouping with other machines (or machine groups) in step 2. This priority rules out the possibility of all machines in one group.

Example 4.5

Consider the initial part–machine given in Fig. 3.11 and illustrate the application of the algorithm proposed by Chow (1992).

Step 1. The commonality scores between machine pairs are computed and are shown in Fig. 4.8.
Step 2. Group machines 1 and 3 with the highest commonality scores.
Step 3. The new machine unit (1,3) $= (1,1,1,0,1,1,0,1)$. The new part–machine matrix is shown in Fig. 4.9(a).
Step 4. Proceed to step 1, since the number of machine units is still not one.
Step 1. The commonality scores for the revised part–machine matrix are shown in Fig. 4.9(b).
Steps 2, 3, 4 and 1. Group machines 2 and 5. The machine unit for this group is (2,5) $= (0,0,1,1,0,1,1,0)$. The revised part–machine matrix and commonality scores between machines are given in Fig. 4.10(a) and (b).
Steps 2, 3, 4 and 1. Group machine unit (1,3) and machine 6 to form a new machine group (1, 3, 6). The part–machine matrix and commonality scores are revised again as in Figs. 4.11(a) and (b).
Steps 2 and 3. Group machine 4 and machine unit (2, 5) to form (2, 5, 4). Revise the part–machine matrix and commonality scores as in Fig. 4.12(a) and (b).

(a)

	1	2	3	4	5	6	7	8
2			1	1		1	1	
1.3	1	1	1		1	1		1
4				1			1	1
5			1	1			1	
6	1	1				1		

(b)

	2	1,3	4	5	6
2	0	14	17	25	9
1,3		0	7	7	23
4			0	18	2
5				0	2
6					0

Fig. 4.9 (a) Part–machine matrix after grouping machines 1 and 3; (b) commonality scores for part–machine matrix of (a).

(a)

	1	2	3	4	5	6	7	8
1,3	1	1	1		1	1		1
4				1			1	1
2,5			1	1		1	1	
6	1	1				1		

(b)

	1,3	4	2,5	6
1,3	0	7	14	23
4		0	17	2
2,5			0	9
6				0

Fig. 4.10 (a) Revised part–machine matrix after grouping machines 2 and 5; (b) commonality scores for revised part–machine matrix of (a).

Step 4. Since the desired number of machine groups is two, stop.

This algorithm generates fewer bottleneck parts, especially when the number of machine cells is greater than or equal to four. This result is based on an empirical study of three data sets for a number of machine groups in the range 2 to 9 and on a comparison with LCC and ALC. It does not guarantee the global minimization of bottleneck parts, but it is the best approach to grouping, say, $(K+2)$ existing machine groups to form $(K+1)$ machine groups (Chow, 1992).

(a)

	1	2	3	4	5	6	7	8
4				1			1	1
2,5			1	1		1	1	
1,3,6	1	1	1		1	1		1

(b)

	4	2,5	1,3,6
4	0	17	7
2,5		0	14
1,3,6			0

Fig. 4.11 (a) Revised part–machine matrix after grouping (1,3) and 6; (b) commonality scores for revised part–machine matrix of (a).

(a)

	1	2	3	4	5	6	7	8
2,5,4			1	1		1	1	1
1,3,6	1	1	1		1	1		1

(b)

	2,5,4	1,3,6
2,5,4	0	21
1,3,6		0

Fig. 4.12 (a) Revised part–machine matrix after grouping 4 and (2,5); (b) commonality scores for revised part–machine matrix of (a).

4.6 EVALUATION OF MACHINE GROUPS

General approach

The level of similarity at which the tree is cut determines the number of machine groups. Determining this level depends on whether the purpose is general or specific. In general, one strategy is to cut the tree at some point within a wide range of the resemblance coefficients for which the number of clusters remains constant, because a wide range indicates that the clusters are well separated in the attribute space (Romesburg, 1984). This means that the decision regarding where to cut the tree is least sensitive to error when the width of the range is largest.

Example 4.6

Table 4.1 summarizes the number of clusters for different ranges of S_{mn} for the tree shown in Fig. 4.7. For this example, it means that forming two machine groups is a good choice, while forming five is a bad choice.

Inter- and intra-group movement

Although the general approach provides a way to determine the number of machine groups, some of the factors which need to be examined prior to determining the number of groups are: the number of inter- and intra-group movements, machine utilization, planning and control factors etc. If the tree is cut at a high similarity value a large number of small machine groups will be identified, while a low similarity value will result in a few machine groups which are large. A large number of small machine groups will lead to an increase in the inter-group movement and a decrease in the intra-group movement. If a small number of large machine groups are identified, the impact will be the opposite. Thus it is important to evaluate the sum of the cost of intra- and inter-group movement for different levels of similarity and identify the machine groups such that the total cost is a minimum. The total intra- and inter-group movement is affected by the location of machine groups and the arrangement of machines within a group. These distances can be estimated using CRAFT (Seiffodini and Wolfe, 1987). However, since the sequence of operations has been ignored so far, and also typically each cell does not consist of many machines, it is reasonable to assume that machines are laid out in a random manner and compute the expected distance a part will travel based on a straight line layout, a rectangle layout or a square layout (McAuley, 1972). The expected distance a part travels between two machines in a group of M machines is:

- $(M+1)/3$ for a straight line;
- $(R+L)/3$ for a rectangle in case of R rows of L machines;
- $2\sqrt{M}/3$ for a square.

This is a reasonable assumption since most layouts follow one of these patterns with passageways between machines and often no diagonal moves are allowed. Thus, if N_j is the number of inter-group journeys for the jth solution, D_j is the total distance for the jth solution, C_1 is the cost of an inter-group journey, and C_2 is the cost per unit distance of an intra-group journey, the best solution is the one which gives minimum

Table 4.1 Number of groups for different ranges of S_{mn}

Number of groups	*Range of S_{mn}*	*Width of range*
6	$0.75 < S_{mn} < 1$	0.25
5	$0.67 < S_{mn} < 0.75$	0.08
4	$0.45 < S_{mn} < 0.67$	0.22
2	$0.093 < S_{mn} < 0.45$	0.357
1	$0.0 < S_{mn} < 0.093$	0.093

cost, i.e. $j^{\min}(N_jC_1 + D_jC_2)$. Also, the solution is not sensitive to the ratio of intra-group and inter-group travel costs, i.e. even if the cost of an inter-group journey varies from four to eight times that of one unit distance covered in an intra-group journey, the solution does not change (McAuley, 1972).

Example 4.7

Consider the dendogram in Fig. 4.16. The five possible solutions are shown in Table 4.2. A good solution is the one with the least total cost of inter- and intra-group travels. The number of inter-group and intra-group travels and the intra-group distances for the five possible solutions are summarized in Table 4.2. For example, in solution 4, the number of intra-group travels for the group (1, 3, 6) is 7 (two each for parts 1 and 2, one each for parts 5, 6 and 8, and none for parts 3, 4 and 7). Similarly the number of intra-group travels for the machine group (4, 2, 5) is 5. For this example, assuming a line layout, the total distance of intra-group travels for this solution is $\{(3+1)/3\} \times 7 + \{(3+1)/3\} \times 5 = 16$. The number of inter-group travels for this solution is 3 (one each for parts 3, 6, and 8). Assuming $C_1 = 10$ and $C_2 = 2$, the total cost for solution 4 is $= (10 \times 3) + (2 \times 16) = 62$. The total cost for each solution is shown in Table 4.3. Solution 4, with the least total cost of 62, and two machine groups, is identified as the best.

Machine duplication

In most practical situations, once the machine groups and parts are identified there are always a few exceptional parts and bottleneck machines. In many cases, there is usually more than one copy of each type of machine. The part–machine matrix does not indicate the existence of such copies. For example, if in Fig. 3.16 there were two copies of machine 4, then one copy can be assigned to the group (3, 1, 6), thus decreasing the inter-group travel of part 8. This can be done without a cost analysis if the load distribution in each group is such that

Table 4.2 Evaluation of the different numbers of groups for inter- and intra-group travels

Solution	*Number of groups*	*Machines in each group*	*Inter-group travels*	*Intra-group travels*	*Intra-group distance*
1	6	(1) (3) (6) (4) (2) (5)	15	0	0
2	5	(1) (3) (6) (4) (2, 5)	12	3	3
3	4	(1, 3) (6) (4) (2, 5)	8	7	7
4	2	(1, 3 6) (4, 2, 5)	3	12	16
5	1	(1, 3, 6, 4, 2, 5)	0	15	35

the requirements of corresponding parts are fully satisfied within the group. The duplication should start with the machine generating the largest number of inter-group moves. If additional copies are not available, however, a machine can be purchased if the associated reduction in inter-group travel cost is greater than the cost of duplication. To determine the way duplication should be carried out, identify the group (other than the parent group) with the largest number of parts processed on the bottleneck machine and determine the number of machines required to make the group independent as follows (Seiffodini, 1989a):

$$N = \frac{\sum_{p=1}^{EP} T_p d_p}{HC(1-P)} \tag{4.7}$$

where N is the number of machines required to make a group independent, T_p is the processing time of part p on a machine (in hours), d_p is the demand for part type p in the planning horizon (weeks), H is the production time (hours) available per week, C is the machine use factor, P is the defective fraction, and EP is the number of exceptional parts produced on the machine.

This analysis will determine the distribution of machines between different machine groups, assuming the machine requirement for the conventional method has been determined. If the required number of machines is an integer, or the fraction is large enough to assign one machine, the assignment of machine(s) should be carried out without cost analysis. This is true because this machine is a part of a general machine requirement of the production schedule in the conventional manufacturing system rather than a requirement for making the group independent. If the required number of machines is a real number, however, an additional machine has to be purchased for the fractional part. This additional machine is required to make the group independent and has to be justified by comparing with the reduction in inter-group material handling cost (a number of other costs are also involved, which cannot be difficult to quantify for this purpose). The reduction in material handling can be estimated by determining the near-optimal layout before and after duplication using a plant layout

Table 4.3 Total travel costs

Solution	*Number of groups*	*Total cost*	
1	6	$(10 \times 15) + (2 \times 0) = 150$	
2	5	$(10 \times 12) + (2 \times 3) = 126$	
3	4	$(10 \times 8) + (2 \times 7) = 94$	
4	2	$(10 \times 3) + (2 \times 16) = 62$	(best solution)
5	1	$(10 \times 0) + (2 \times 35) = 70$	

algorithm such as CRAFT (Seifoddini and Wolfe, 1987; Seifoddini, 1989a). If the saving in inter-group material handling cost equals or exceeds the duplication cost, the purchase of a new machine is justified. It is, however, recommended that other factors such as setup costs and cost savings due to better scheduling be considered in the decision-making process. Moreover, if the fraction is very small, other alternatives such as subcontracting or generating an alternate process plan should be considered prior to evaluating the duplication alternative.

4.7 PARTS ALLOCATION

To complete the cell formation, the parts need to be allocated to the machine groups identified. This can be done in one of the following ways:

1. Allocate each part to the machine group which can perform the maximum number of operations. If a machine group is not assigned any parts, assign these machines to the groups where they can perform the maximum number of operations.
2. One of the algorithms such as ROC or the DCA can be performed on the part columns alone for the machine groups obtained.
3. Use the clustering algorithm to construct the part dendogram by defining the similarity between pairs of parts p and q as

$$S_{pq} = a/(a+b+c), \quad 0.0 \leqslant S_{pq} \leqslant 1.0 \tag{4.8}$$

where the two-by-two table is shown as Fig. 4.13 and a is the number of machines processing both parts, b is the number of machines processing part p and not q, c is the number of machines processing part q and not p, and d is the number of machines processing either part.

It is important to note that when the two groups are combined, the ordering of machines within the new group should retain the ordering of the machines in the two groups. This ordering also applies to parts. From this final matrix the partition can be performed manually. A number of different partitions can be selected and one or more of the performance measures discussed in Chapter 3 can be used to identify a good solution. This approach is especially useful in the absence of

		Part q 1	0
Part p	1	a	b
	0	c	d

Fig. 4.13 2 × 2 parts table.

information on the machines layout, cost etc. Also, the problem of chaining can be avoided.

Example 4.8

Illustrate the approach to cell formation applying ALC on parts and machines. The Jaccards similarity matrix for parts is given in Fig. 4.14. The dendogram for parts and machines and the ordering is shown in Fig. 4.15.

Example 4.9

The dendogram for parts based on SLC and CLC is shown in Fig. 4.16.

4.8 GROUPABILITY OF DATA

Although a number of algorithms have been proposed for block diagonalization and clustering, it may so happen that the matrix itself may not be amenable to such groupability, however good the algorithm is. Thus, it is important to characterize the factors which affect this groupability. Chandrasekaran and Rajagopalan (1989), based on an experimental study of a few well-structured to ill-structured matrices, presented the following as a set of possibilities:

1. Whatever the similarity or dissimilarity used for the purpose of block diagonalization, the Jaccards similarity coefficient S was found to be most suitable for analysing the groupability of matrices.
2. As the matrix becomes ill-structured, the spread (standard deviation σ_S) of the pairwise similarities decreases and decreases the grouping efficiency.
3. The final grouping efficiency is strongly related to the standard deviation σ_S and the average $\bar{s}$ of the pairwise similarities, although the relation with standard deviation is more pronounced in terms of absolute values.

	1	2	3	4	5	6	7	8
1	0	1	0.2	0	0.67	0.5	0	0.5
2		0	0.2	0	0.67	0.5	0	0.5
3			0	0.5	0.25	0.5	0.5	0.2
4				0	0	0.2	1	0.2
5					0	0.25	0	0.67
6						0	0.2	0.2
7							0	0.2
8								0

Fig. 4.14 Jaccards similarity matrix for parts.

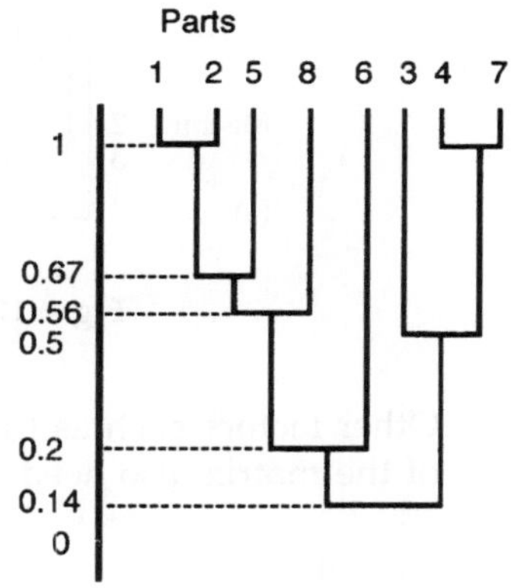

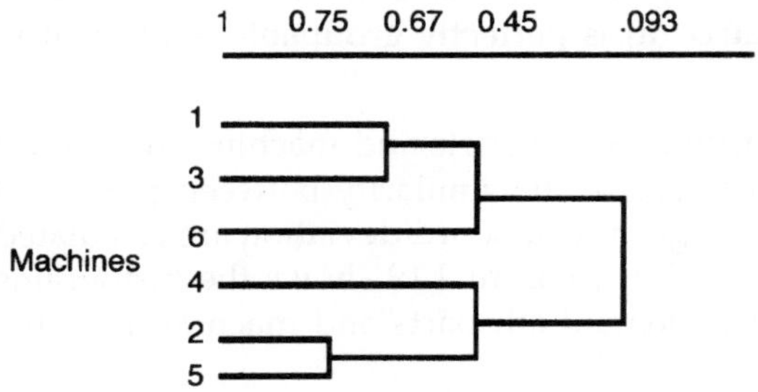

	1	2	5	8	6	3	4	7
1	1	1	1	1				
3	1	1	1	1	1	1		
6	1	1			1			
4				1			1	1
2					1	1	1	1
5						1	1	1

Fig. 4.15 Part and machine reordering using ALC.

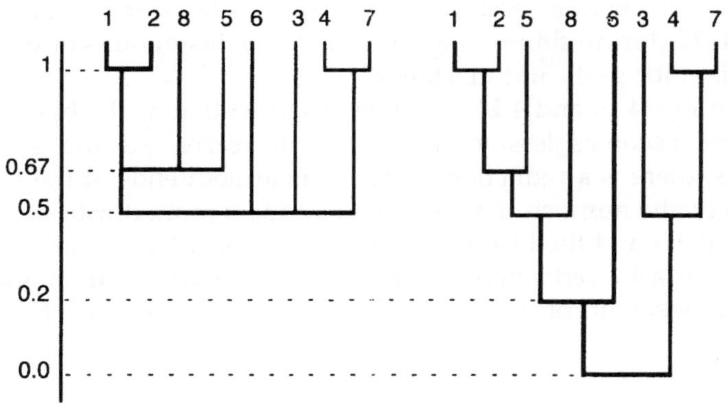

Fig. 4.16 Dendogram for parts using SLC and CLC.

4. For matrices encountered in cell formation, it can be concluded that the working range of σ_S is between 0.2 and 0.35. Data are ill-structured, too sparse or too dense if they fall outside this range. The size of matrices considered in this study was 40 * 24.

(a)

Machine \ Part	1	2	3	4
1	1	1		
2	1	1		
3			1	1
4			1	1

(b)

Machine \ Part	1	2	3	4
1	1		1	
2	1	1	1	1
3	1			1
4		1	1	1

Fig. 4.17 Matrices for Example 4.10.

5. Other factors such as the number of machines, parts and the density of the matrix also need to be considered for a more accurate picture.

Example 4.10

Consider the two matrices in Fig. 4.17: the density of matrix (a) is 0.5 and matrix (b) is 0.6875; matrix (a) is perfectly groupable while matrix (b) is not.

The pairwise Jaccards similarity between parts and machines for matrix (a) is shown in Fig. 4.18. In this case the similarity between parts and machines is identical. The average and standard deviation are calculated to be $\bar{s} = 2/6 = 0.333$ and $\sigma_S = 0.5163$. Figure 4.19 shows the histograms of the Jaccards similarity coefficients (both parts and machines) for the matrix of Fig. 4.17(a).

The pairwise Jaccards similarities for parts and machines for the matrix of Fig. 4.17(b) are shown in Fig. 4.20(a) and (b). The average and standard deviations are: $\bar{s} = 0.473$, $\sigma_s = 0.1889$ (for parts); $\bar{s} = 0.347$, $\sigma_s = 0.1277$ (for machines). Figure 4.21 shows histograms of the Jaccards coefficient for parts and machines.

From Fig. 4.19 and 4.21 it can be observed that as the block diagonal structure becomes less feasible so σ_s decreases. As the groupability reduces, there is a reduction of elements at both ends of the histogram. However, the number of similar pairs reduces more drastically than dissimilar pairs and the histograms tends to consolidate towards zero. This causes a drastic reduction in the spread of the distribution and a zero-ward movement of the average (Chandrasekaran and Rajagopalan, 1989).

	1	2	3	4
1	–	1	0	0
2	–	–	0	0
3			–	1
4				–

Fig. 4.18 Pairwise Jaccards similarity between parts and machines for Fig. 4.17(a)

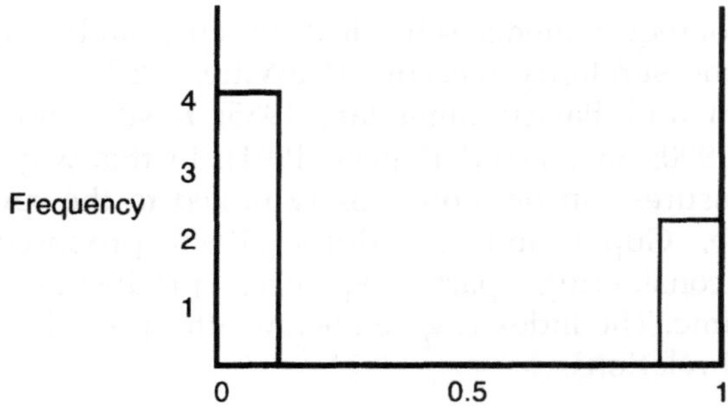

Fig. 4.19 Histogram of Jaccards coefficient for Fig. 4.17(a) (part or machine).

(a) Part

	1	2	3	4
1	–	0.25	0.5	0.5
2	–	–	0.67	0.67
3			–	0.25
4				–

(b) Machine

	1	2	3	4
1	–	0.5	0.33	0.25
2		–	0.5	0.25
3			–	0.25
4				–

Fig. 4.20 Pairwise Jaccards similarity: (a) for parts; (b) for machines.

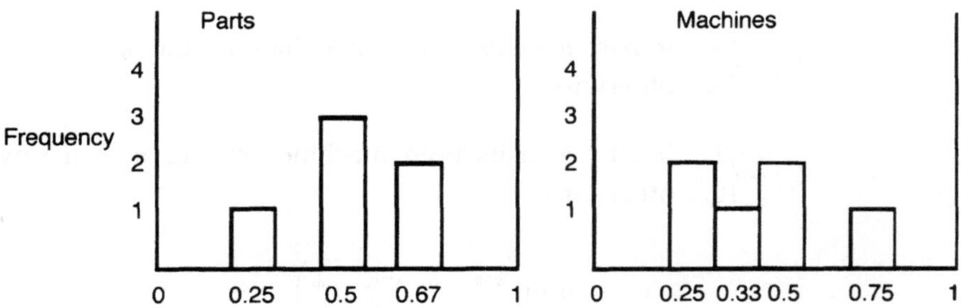

Fig. 4.21 Histogram of Jaccards coefficient for Fig. 4.17(b).

4.9 RELATED DEVELOPMENTS

In two articles Shafer and Rogers (1993a, b) reviewed the different similarity and distance measures used in cellular manufacturing. The Jaccards similarity introduced in this chapter is the simplest form of measure requiring the information provided in the part–machine matrix. Other manufacturing features such as part volume, part sequence, tool requirements, setup features etc., can be considered while computing the similarity measure (DeWitte, 1980; Mosier and Taube, 1985; Selvam and Balasubramanian, 1985; Kasilingam and Lashkari, 1989; Tam, 1990; Shafer and Rogers, 1991). In this way similarity and distance measures can be more closely linked to the specific situation. For example, Gupta and Seifoddini (1990) proposed a similarity coefficient considering part sequence, production volume and processing time. The index is given below after providing the necessary notation and relations.

Notation

S_{mn} similarity coefficient between machines m and n
m_k production volume for part k
n_k number of times part k visits both machines in a row (or succession)
η_m^p number of trips part p makes to machine m
η_n^p number of trips part p makes to machine n
t_m^{po} unit operation time for part p on machine m during the oth visit
t_n^{po} unit operation time for part p on machine n during the oth visit
t_p^{mn} ratio of total smaller unit operation time to the larger unit operation time for machine pair mn, for part p during visits to machine m and n.

$$X_p = \begin{cases} 1, & \text{if part } p \text{ visits both machines } m \text{ and } n \\ 0, & \text{otherwise} \end{cases}$$

$$Y_p = \begin{cases} 1, & \text{if part } p \text{ visits either machines } m \text{ and } n \\ 0, & \text{otherwise} \end{cases}$$

$$Z_{po} = \begin{cases} 1, & \text{if part } p \text{ visits both machines } m \text{ and } n \text{ in a row} \\ 0, & \text{otherwise} \end{cases}$$

$$S_{mn} = \frac{\min\left(\sum_{0=1}^{\eta_m^p} t_m^{po}, \sum_{0=1}^{\eta_n^p} t_n^{po}\right)}{\max\left(\sum_{0=1}^{\eta_m^p} t_m^{po}, \sum_{0=1}^{\eta_n^p} t_n^{po}\right)}$$

The new similarity coefficient (taken from Gupta and Seifoddini (1990)) can be defined as:

$$S_{mn} = \frac{\sum_{p=1}^{P}\left[X_p t_p^{mn} + \sum_{0=1}^{\eta_p} Z_{po}\right] m_p}{\sum_{p=1}^{P}\left[X_p t_p^{mn} + \sum_{0=1}^{\eta_p} Z_{po} + Y_p\right] m_p}$$

This measure computes the similarity as a weighted term for each part visiting at least one of the two machines. The weighting is determined by the average production volume, part sequence and unit processing time for each operation. Thus, a high-volume part that is processed by a pair of machines will contribute more towards their similarity than a low-volume part. Also, the product of production volume and unit operation time determines the workload for a part. Higher similarity values are indirectly assigned to those pairs of machines which process parts with larger workload. The sequence is considered by giving higher priority to those machines which need more handling. Once these measures are computed for all machine pairs, the clustering algorithms discussed in this chapter can be used to identify the machine groups. Abundant research literature is available on traditional clustering procedures applied to a variety of problems. However, in the context of cell formation, it is interesting to note that the research devoted to machine grouping procedures outnumbers the part grouping procedures by almost two to one (Shafer and Rogers, 1993a, b). For a comparison of the applications of clustering methods refer to Mosier (1989) and Shafer and Meredith (1990).

4.10 SUMMARY

The clustering methods introduced in this chapter adopt a sequential approach to cell formation. Once the part–machine matrix is available, a suitable measure of similarity or dissimilarity between machines is defined. This is followed by the selection of a clustering method to result in the dendogram. Depending on the situation, the user decides the number of machine groups using one of the criteria listed in this chapter. Subsequently the part allocation to these machine groups is obtained.

The Jaccards similarity and commonality measures discussed here require only information provided in the part–machine matrix. However, procedures using a similarity coefficient method are flexible to consider manufacturing features such as part volumes, part sequence, processing times, setup times etc., while computing the similarity measure.

The clustering algorithms remain unaffected by the definition of the similarity measure. In fact, the availability of commercial software packages for the clustering algorithms discussed in this chapter makes these procedures more attractive than the matrix manipulation algorithms. However, all the methods discussed in Chapters 3 and 4 are heuristics and are data dependent, i.e. the input data could be structured (a pure block diagonal form exists) or unstructured (however good an algorithm, the data cannot be decomposed to a pure diagonal form with non-overlapping elements). Thus, it is useful to know about the nature of an input matrix before using any of these heuristic procedures.

The Jaccards similarity has been found most suitable for analysing the groupability of matrices. The standard deviation σ_S and the average $\bar{s}$ of the pairwise similarities are strongly related to the grouping efficiency of a matrix. Thus, an input matrix which is ill-structured has low values of σ_S and $\bar{s}$ of the pairwise similarities. However, further research is warranted to understand the ability of different algorithms to provide a good partition in relation to factors which affect the grouping efficiency of input matrices. This would assist the user in selecting the best heuristic procedure for a given situation after identifying the nature of the input matrix. For example, if the input matrix is perfectly groupable, then the modified CIA, which is the most efficient algorithm, can be selected and applied to the data.

PROBLEMS

4.1 What is the significance of similarity or dissimilarity in clustering machines?

4.2 Consider the part–machine matrix of Fig. 4.22. Apply SLC, CLC and ALC to machines using the Jaccards similarity as a measure. Draw the dendograms for each case and compare. Based on the general approach, how would you cut the dendogram and identify the machine groups? What do you observe is the advantage of one method over the other? If the machines within a cell are arranged in a straight line, the cost of an intra-cell move per unit distance is $5 and the inter-cell cost is $15, what is the most economical number of machine groups? For these machine groups determine the part allocation. What are the different options available for dealing with exceptional parts and bottleneck machines? Under what circumstances do you consider machine duplication as a viable option for dealing with bottleneck machines?

4.3 How does the commonality measure differ from the Jaccards similarity measure?

4.4 Apply LCC to the data in Q 4.2.

4.5 What factors influence the groupability of a part–machine matrix? Discuss the use of standard deviation as a means to classify matrices as being well-structured or ill-structured.

	p1	p2	p3	p4	p5
m1		1	1		1
m2	1	1		1	
m3	1				1
m4		1	1	1	1
m5	1	1	1		
m6			1	1	1
m7	1		1		1
m8	1	1	1		1

Fig. 4.22 Part–machine matrix for Q4.2.

REFERENCES

Chandrasekaran, M. P. and Rajagopalan, R. (1989) Groupability: an analysis of the properties of binary data matrices for group technology. *International Journal of Production Research*, **27**(7), 1035–52.

Chow, W. S. (1991) A note on a linear cell clustering algorithm. *International Journel of Production Research*, **29**(1), 215–16.

Chow, W. S. (1992) Efficient clustering and knowledge based approach for solving cellular manufacturing problems. Univ. Manitoba, Canada. Ph.D. dissertation.

De Witte, J. (1980) The use of similarity coefficients in production flow analysis. *International Journal of Production Research*, **18**, 503–14.

Gupta, T. and Seifoddini, H. (1990) Production data based similarity coefficient for machine-component grouping decisions in the design of a cellular manufacturing system. *International Journal of Production Research*, **28**(7), 1247–69.

Kasilingam, R. G. and Lashkari, R. S. (1989) The cell formation problem in cellular manufacturing systems – a sequential modeling approach. *Computers and Industrial Engineering*, **16**, 469–76.

McAuley, J. (1972) Machine grouping for efficient production. *The Production Engineer*, **51**(2), 53–7.

Mosier, C. T. (1989) An experiment investigating the application of clustering procedures and similarity coefficients to the GT machine cell formation problem. *International Journal of Production Research*, **27**(10), 1811–35.

Mosier, C. T. and Taube, L. (1985) Weighted similarity measure heuristics for the group technology machine clustering problem, *Omega*, **13**, 577–9.

Romesburg, H. C. (1984) *Cluster Analysis for Researchers*, Lifetime Learning Publications, Belmont, CA.

Seifoddini, H. (1989a) Duplication process in machine cells formation in group technology. *IIE Transactions*, **21**(4), 382–8.

Seifoddini, H. (1989b) A note on the similarity coefficient method and the problem of improper machine assignment in group technology applications. *International Journal of Production Research*, **27**(7), 1161–5.

Seifoddini, H. and Wolfe, P. M. (1987) Selection of a threshold value based on material handling cost in machine–component grouping. *IIE Transactions*, **19**(3), 266–70.

Selvam, R. P. and Balasubramanian, K. N. (1985) Algorithmic grouping of operation sequences. *Engineering Costs and Production Economics*, **9**, 125–34.

Shafer, S. M. and Meredith, J. R. (1990) A comparison of selected manufacturing cell formation techniques. *International Journal of Production Research*, **28**(4), 661–73.

Shafer, S. M. and Rogers, D. F. (1991) A goal programming approach to cell formation problem. *Journal of Operations Management*, **10**, 28–43.

Shafer, S. M. and Rogers, D. F. (1993a) Similarity and distance measures for cellular manufacturing, Part 1: a survey. *International Journal of Production Research*, **31**(5), 1133–42.

Shafer, S. M. and Rogers, D. F. (1993b) Similarity and distance measures for cellular manufacturing, Part 2: an extension and comparison. *International Journal of Production Research*, **31**(6), 1315–26.

Tam, K. Y. (1990) An operation sequence based similarity coefficient for part families formations. *Journal of Manufacturing Systems*, **9**(1), 55–68.

Wei, J. C. and Kern, G. M. (1989) Commonality analysis: a linear cell clustering algorithm for group technology. *International Journal of Production Research*, **27**(12), 2053–62.

Wei, J. C. and Kern, G. M. (1991) Reply to 'A note on a linear cell clustering algorithm'. *International Journal of Production Research*, **29**(1), 217–18.

CHAPTER FIVE

Mathematical programming and graph theoretic methods for cell formation

The algorithmic procedures for cell formation discussed so far are heuristics. As discussed, these procedures are affected by the nature of input data and the initial matrix and do not necessarily provide a good partition, even if one is possible. Thus, there is a need to develop mathematical models which can provide optimal solutions. The models provide a basis for comparison with the heuristics. The structure of the model thus developed also assists the researcher in suggesting efficient solution schemes. Moreover, the heuristics can be used as a starting point to drive an optimal algorithm towards searching for better, or even optimal solutions while saving on a great deal of computer time (Wei and Gaither, 1990).

The number of cells, parts and machines in each cell is determined subsequently by the application of the matrix manipulation algorithms and clustering algorithms. This, in one sense, allows the user to identify natural groups. However, in most mathematical models this information is an input. Several factors affect these parameters: physical shopfloor layout, labor-related issues, the need for uniform cell size, production control issues etc.

This chapter presents some mathematical models which can be used for part family formation and/or machine grouping. Depending on the model and objective, the user will adopt a sequential or simultaneous approach to cell formation. The impact of considering alternative process plans and additional machine copies if available will be discussed. A mathematical model considering these aspects is also presented. Finally, the major algorithms discussed in Chapters 3 to 5 will be reviewed.

5.1 P-MEDIAN MODEL

Kusiak (1987) proposed the p-median model to identify part families. This was the first approach to forming part families using mathematical

Machines \ Parts	1	2	3	4	5	6	7	8
1	1	1			1			1
2			1	1		1	1	
3	1	1	1		1	1		1
4				1			1	1
5			1	1			1	
6	1	1				1		

Fig. 5.1 Initial part–machine matrix for example 5.1.

programming. The mathematical model remains the same as in Chapter 2, except here we consider the maximization of similarity instead of minimizing distance. The number of medians f is a given parameter in the model. The model selects f medians and assigns the remaining parts to these medians such that the sum of similarity in each part family is maximized. Similarity between two parts is defined as the number of machines the two parts have in common, i.e.

$$S_{pq} = \sum_{p=1}^{P} \delta(a_{pm,} a_{pn}) \tag{5.1}$$

where

$$\delta(a_{pm,} a_{pn}) = \begin{cases} 1 & a_{pm} = a_{pn} \\ 0 & \text{otherwise} \end{cases}$$

Example 5.1

Consider the matrix of eight parts and six machines given in Fig. 5.1. The similarity between parts calculated using equation 5.1 is given in Fig. 5.2. By considering the similarity between parts given in Fig. 5.2, if the p-median model is solved to obtain two part families: $X_{11} = X_{21} = X_{51} = X_{61} = X_{81} = 1$; $X_{34} = X_{44} = X_{74} = 1$ and all other $X_{pq} = 0$. Thus, one part family consists of parts $\{1, 2, 5, 6, 8\}$ and the other part family consists of parts $\{3, 4, 7\}$. The median parts are 1 and 4 and the objective value is 41.

Limitations of the p-median model

1. This procedure identifies only the part families; an additional procedure is needed to identify the machine groups.
2. The correct value of f to identify a good block diagonal is not known. Moreover, the best value of f need not correspond to the highest

	1	2	3	4	5	6	7	8
1	–	6	2	0	5	4	0	4
2		–	2	0	5	4	0	4
3			–	4	3	4	4	2
4				–	1	2	6	2
5					–	3	1	5
6						–	2	2
7							–	2
8								–

Fig. 5.2 Similarity between parts.

value of the objective function. Thus, one has to experiment with the value of f.

5.2 ASSIGNMENT MODEL

To avoid the problem of determining the optimal value of f, Srinivasan, Narendran and Mahadevan (1990) proposed an assignment model for the part families and machine grouping problem. They provided a sequential procedure to identify machine groups followed by identification of part families. The objective of the assignment model is to maximize the similarity. The definition of similarity is as in equation 5.1. On solving the model, sub-tours (closed loops) are identified in the solution. Each identified closed loop forms the basis for grouping parts and machines. The proposed algorithm consists of two stages. If the matrix is mutually separable, the procedure stops after stage 1. However, if the solution will result in exceptional elements, stage 2 is activated, where part families are assigned to machine groups in such a way that will result in minimum exceptional elements and voids. The assignment model for part family formation and machine grouping is given below, where p,q are indexes for parts and m,n are indexes for machines, and the following relations hold:

$$X_{pq} = \begin{cases} 1 & \text{if part } p \text{ and } q \text{ are connected} \\ 0 & \text{otherwise} \end{cases}$$

$$Y_{mn} = \begin{cases} 1 & \text{if machine } m \text{ and } n \text{ are connected} \\ 0 & \text{otherwise} \end{cases}$$

Part family model

Maximize

$$\sum_{p=1}^{P}\sum_{q=1}^{P} S_{pq} X_{pq}$$

subject to:

$$\sum_{q=1}^{P} X_{pq} = 1, \quad \forall p \tag{5.2}$$

$$\sum_{p=1}^{P} X_{pq} = 1, \quad \forall q \tag{5.3}$$

$$X_{pq} = 0/1, \quad \forall p,q \tag{5.4}$$

Constraints 5.2 and 5.3 ensure that each part has a follower and a predecessor to form a closed loop. The integer nature of the decision variables is identified by constraints 5.4.

Machine grouping model

Maximize

$$\sum_{m=1}^{M}\sum_{n=1}^{M} S_{mn}\, Y_{mn}$$

subject to:

$$\sum_{n=1}^{M} Y_{mn} = 1, \quad \forall m \tag{5.5}$$

$$\sum_{m=1}^{M} Y_{mn} = 1, \quad \forall n \tag{5.6}$$

$$Y_{mn} = 0/1, \quad \forall m,n \tag{5.7}$$

Constraints 5.5 to 5.7 correspond to constraints 5.2 to 5.4, respectively.

Algorithm

Stage 1

Step 1. Compute similarity coefficients S_{mn} between machines.
Step 2. Use the coefficients S_{mn} as an input to the assignment model and solve it for maximization (machine grouping model).
Step 3. Identify all closed loops. Each closed loop forms a machine group.
Step 4. List all the parts that visit each group.

Step 5. Scan the list of parts visiting each group. Whenever the part family for a machine group is a subset of another, merge them into one. Repeat this process until no further grouping is possible.
Step 6. If the part families are disjoint, stop; else, proceed to stage 2.

Stage 2

Step 7. Repeat steps 1 to 3 to identify part families (use part grouping model in step 2).
Step 8. Assign a part family f to a machine group g on which the maximum number of operations can be performed. Repeat this procedure to assign all part families. Ties can be broken arbitrarily.
Step 9. If there is any machine group which has no part families assigned to it, merge it with an existing group where it can perform the maximum number of operations. Repeat this procedure until all machine groups are non-empty.
Step 10. Merge two groups g and h and their part families if the number of voids created by the merger is not more than the number of exceptional elements eliminated by the merger. Stop when no more mergers are possible.

Example 5.2

Consider the part–machine matrix in Fig. 5.1 and illustrate the assignment model approach to identifying part families and machine groups.

Stage 1

Step 1. Compute the similarity matrix for machines (Fig. 5.3).
Step 2. Using the similarity matrix, solving the assignment model gives $Y_{16} = Y_{63} = Y_{31} = 1$; $Y_{25} = Y_{54} = Y_{42} = 1$. The objective value is 34.
Step 3. The closed machine loops (groups) are (1–6–3) and (2–5–4).
Step 4. The parts which visit each group are given in Table 5.1.
Step 5. No merging is possible.

	1	2	3	4	5	6
1	–	0	6	3	1	5
2		–	2	5	7	3
3			–	1	1	5
4				–	6	2
5					–	2
6						–

Fig. 5.3 Similarity between machines.

Table 5.1 Parts visiting each group

Machine group	*Machines*	*Parts*
1	1,6,3	1,2,3,5,6,8
2	2,5,4	3,4,6,7,8

Table 5.2 Four part families

Part family	*Parts*
1	1,2
2	3,6
3	4,7
4	5,8

Step 6. The part families are not disjoint, since parts 3,6 and 8 are visiting both cells. Proceed to stage 2.

Stage 2

Step 7. Solve the assignment model for forming part families using the similarity measures given in Fig. 5.2. On solving, the following closed loops are identified: $X_{12} = X_{21} = 1$; $X_{36} = X_{63} = 1$; $X_{47} = X_{74} = 1$; $X_{58} = X_{85} = 1$, i.e. four part families are formed (Table 5.2.)
Step 8. Assign part family f to the group which can perform the maximum number of operations. The number of operations required, and which can be performed in each group for each part family, are given in Table 5.3. Thus, assign PF1 to MG1, PF2 to either group, say MG 2, PF3 to MG2 and PF4 to MG1. The two machine groups and part families are given in Table 5.4.
Step 9. Since each machine group has a part family assigned to it, this step is not required.
Step 10. There are four exceptional elements (1s in bold) and five voids (stars) with the current partition, as shown in Fig. 5.4. If the two groups are merged, 21 additional voids (the 0s) are created, which is greater than the number of exceptional elements, hence do not merge.

This approach was reported to be superior both in terms of quality of solution and computational time on a number of examples in comparison with the p-median model. However, in the above problem if part 6 was assigned to the machine group (1,3,6) it would lead to identification of better groups. This problem arises due to grouping of parts before assigning them to machine groups. Srinivasan and Narendran (1991) developed an iterative procedure called GRAFICS to overcome this limitation.

Table 5.3 Operations on part families

Part family	*Machine group*	
	1(1,3,6)	2(2,4,5)
1(1,2)	6/6	0/6
2(3,6)	3/6	3/6
3(4,7)	0/6	6/6
4(5,8)	4/5	1/5

Table 5.4 Assigning part families

Group	*Machines*	*Parts*
1	1,3,6	1,2,5,8
2	2,5,4	3,6,4,7

Machines \ Parts	1	2	5	8	3	6	4	7
1	1	1	1	*	0	0	0	0
3	1	1	1	1	**1**	**1**	0	0
6	1	1	*	*	0	**1**	0	0
2	0	0	0	0	1	1	1	1
5	0	0	0	0	1	*	1	1
4	0	0	0	**1**	*	*	1	1

Fig. 5.4 Resulting partition in step 10.

5.3 QUADRATIC PROGRAMMING MODEL

The clustering algorithms and p-median model minimize the distance or maximize the similarity between parts by considering the family (group) mean or median. However, the parts within a family interact with each other. Therefore, it becomes important to account for the total family (group) interaction. Further, one should be able to restrict the number of families (groups) and family (group) sizes. Kusiak, Vanelli and Kumar (1986) proposed a quadratic programming model for this purpose. They proposed solving this model by an eigenvector-based algorithm. However, it can be solved by linearizing the objective. In this model f is the index for the part family and F_f is the maximum number of parts in part family f, and

$$X_{pf} = \begin{cases} 1 & \text{if part } p \text{ is assigned to part family } f \\ 0 & \text{otherwise} \end{cases}$$

Part family model

Maximize

$$\sum_{p=1}^{P-1} \sum_{q=p+1}^{P} \sum_{f=1}^{F} S_{pq} X_{pf} X_{qf}$$

subject to:

$$\sum_{f=1}^{F} X_{pf} = 1, \quad \forall p \tag{5.8}$$

$$\sum_{p=1}^{P} X_{pf} \leqslant F_{f,} \quad \forall f \tag{5.9}$$

$$X_{pf} = 0/1, \quad \forall\, p,f \tag{5.10}$$

Constraints 5.8 ensure that each part belongs to exactly one part family. Constraints 5.9 guarantee that part family f does not contain more than F_f parts. The integrality restrictions are imposed by constraints 5.10. The above model can be solved by linearizing the non-linear terms in the objective.

Example 5.3

Using the similarity values given in Fig. 5.2, the above model was solved for $F_1 = F_2 = 4$. The solution to the linear model identifies parts 3, 4, 6 and 7 in part family 1 and parts 1, 2, 5 and 8 in part family 2. The objective value is 51, which is the sum of all interactions of parts within each family. This solution is the same as obtained using the assignment model. To illustrate the impact of values given to $F_{f,}$ the model was solved for $F_1 = 5$, $F_2 = 3$. The objective value in this case is 56 and the part families are identical to those obtained using the p-median model. Thus, the values of F_f significantly affect the part family formation. The maximum objective value is obtained when all parts are in one family.

5.4 GRAPH THEORETIC MODELS

The part–machine matrix $[a_{pm}]$ can also be represented as a graph formulation. Depending on the representation of nodes and edges, three types of graph can be used (Kusiak and Chow, 1988): bipartite graph, transition graph or boundary graph.

Bipartite graph

Instead of performing row and column operations to obtain a block diagonal matrix, here we look equivalently at the decomposition of networks. The problem is formulated as a k-decomposition problem in graph theoretic terms. In a bipartite graph, one set of nodes represents the parts and the other the machines. The edges (arcs) between the two sets of nodes represent the requirement for machine m for part p. A k-decomposition is obtained by deleting edges to obtain k disconnected graphs. The parameter k is equivalent to p in the median formulation. Mathematically the model is the same as the quadratic programming model except the variable X_{pf} is defined as follows ($k = f$):

$$X_{pf} = \begin{cases} 1 & \text{if node } p \text{ is assigned to part family } f \\ 0 & \text{otherwise} \end{cases}$$

It is important to note, however, that node $p(q)$ includes all the nodes corresponding to parts and machines. Thus, if there are five parts and four machines, a total of nine nodes have to be considered. Thus, unlike the quadratic programming model, which identifies only the part families, this model simultaneously identifies the part families and machine groups. Kumar, Kusiak and Vanelli (1986) proposed the quadratic programming model with the objective of maximizing the production flow between machines in each sub-graph. Thus, the coefficient d_{pq} denotes the volume of part p processed on machine q. This is equivalent to minimizing the sum of interdependencies of the k weighted sub-graphs (part families).

To illustrate the bipartite graph, consider the part–machine matrix in Fig. 3.1. The graph is shown in Fig. 5.5. The objective of the model is to determine optimally the edge(s) to be cut to make the graph into two disjoint sub-graphs. For example, if the edge connecting part 3 and machine 1 is cut, two disjoint sub-graphs are identified, as shown in Fig. 5.6.

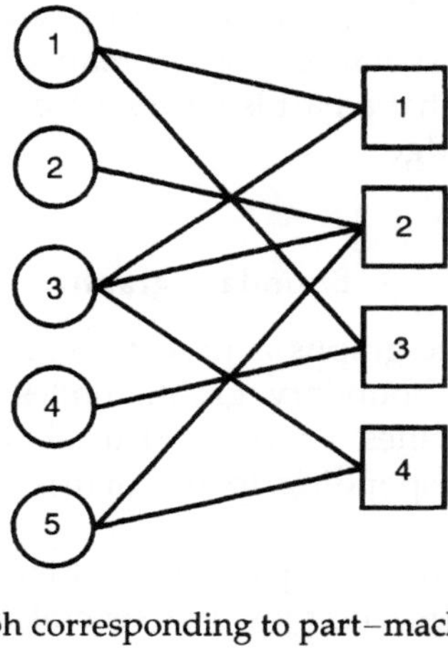

Fig. 5.5 Bipartite graph corresponding to part–machine matrix in Fig. 3.1.

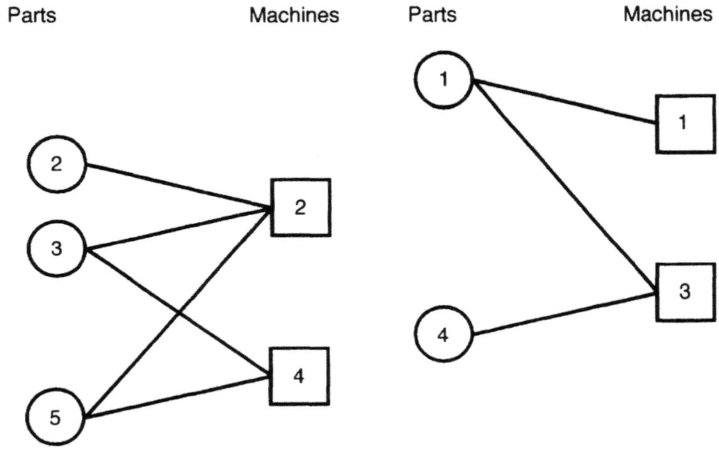

Fig. 5.6 Two disjoint bipartite graphs.

Transition graph

In a transition graph a part (machine) is represented by a node while a machine (part) is represented by an edge. Song and Hitomi (1992) adopted this approach to group machines and to determine the number of cells and cell size, given an upper bound on both. The nodes in this case represent the machines, and two nodes are connected by an edge if d_{mn}, the total number of parts which need these two machines, exists. The objective of the model is to maximize the total number of parts produced within each group, thus minimizing the inter-cell part flows. This is again a quadratic programming problem which decides X_{mg}, i.e. if machine m is assigned to group g or not. The numbers on the arcs denote the number of parts flowing between these two machines. The objective is to divide the machines into g groups (k sub-graphs). A transition graph representation for the matrix in Fig. 3.1 is shown in Fig. 5.7. It is assumed that a part is represented as a node and a machine is represented by an edge.

Boundary graph

A hierarchy of bipartite graphs is used to represent a boundary graph. At each level of the boundary graph, nodes of the bipartite graph represent either machines or parts (Kusiak and Chow, 1988). The boundary graph corresponding to the matrix in Fig. 3.1 is shown in Fig. 5.8.

Determining the bottleneck part or machine in a graph to identify disjoint graphs is rather complex and several authors have addressed

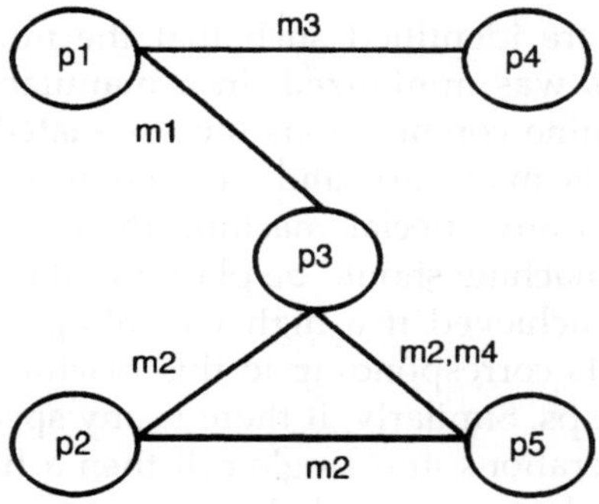

Fig. 5.7 Transition graph corresponding to part–machine matrix in Fig. 3.1.

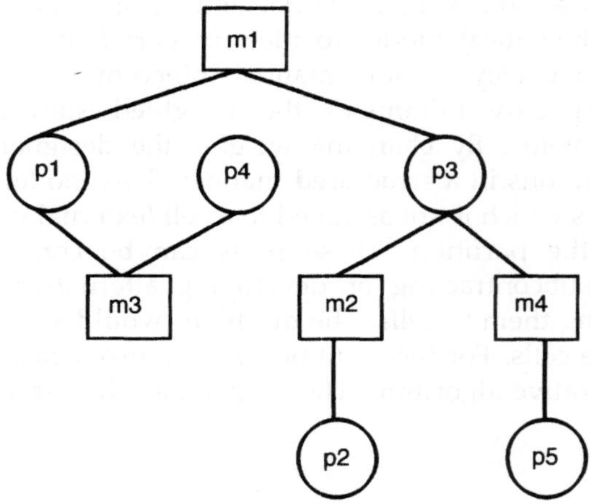

Fig. 5.8 Boundary graph for part–machine matrix in Fig. 3.1.

this problem. Lee, Voght and Mickle (1982) developed a heuristic algorithm to detect the bottleneck parts/machines. This algorithm was further extended by Vannelli and Kumar (1986). A few other graph-based approaches include Rajagopalan and Batra (1975), Vohra *et al.* (1990) and Wu and Salvendy (1993).

5.5 NONLINEAR MODEL AND THE ASSIGNMENT ALLOCATION ALGORITHM (AAA)

The clustering techniques and mathematical models discussed so far consider indirect measures such as similarity/dissimilarity, bond energy, ranking etc., to obtain a block diagonal form. Part families and machine groups were identified such that the number of exceptional elements and voids was minimized. In a manufacturing situation, for different part/machine combinations the associated costs of voids and exceptional elements may vary and in general are not the same. For example, if there is any special machine then all the parts requiring processing on this machine should be placed in the small cell (Burbidge, 1993). This can be achieved if a high weighting value is given to the exceptional elements corresponding to this machine for all parts, while identifying the groups. Similarly, if there is any special part that should complete all its operations in a single cell then a high weighting value should be given to the exceptional elements corresponding to this part. This shows that there is a need to consider the importance of voids and exceptional elements explicitly.

The procedures discussed so far decouple the cell formation and cell evaluation procedure. Adil, Rajamani and Strong (1993a) proposed a nonlinear mathematical model to identify part families and machine groups simultaneously without manual intervention. The objective of the model explicitly minimizes the weighted sum of exceptional elements and voids. By changing weights the designer can generate alternative solutions in a structured manner. This model also identifies parts/machines which if not assigned to a cell (external parts/machines) can enhance the partition. These parts can be considered to have potential for subcontracting or developing alternative process plans before allocating them to cells. The machines would serve as a common resource to the cells. For the solution of large problems, they proposed an efficient iterative algorithm. The model and algorithm are discussed below.

Simultaneous grouping model

Minimize

$$w \cdot \sum_{c=1}^{C} \sum_{p=1}^{P} \sum_{m=1}^{M} a_{pm} \cdot X_{pc}(1 - Y_{mc}) +$$

$$(1 - w) \cdot \sum_{c=1}^{C} \sum_{p=1}^{P} \sum_{m=1}^{M} (1 - a_{pm}) X_{pc} \cdot Y_{mc}$$

subject to:

$$\sum_{c=1}^{C} X_{pc} = 1, \quad \forall p \tag{5.11}$$

$$\sum_{c=1}^{C} Y_{mc} = 1, \quad \forall m \tag{5.12}$$

$$X_{pc,} Y_{mc} = 0/1, \quad \forall p,m,c \tag{5.13}$$

where c is the cell index and

$$a_{pm} = \begin{cases} 1 & \text{if part p requires processing on machine } m \\ 0 & \text{otherwise} \end{cases}$$

$$X_{pc} = \begin{cases} 1 & \text{if part } p \text{ is allocated to cell } c \\ 0 & \text{otherwise} \end{cases}$$

$$Y_{mc} = \begin{cases} 1 & \text{if machine } m \text{ is assigned to cell } c \\ 0 & \text{otherwise} \end{cases}$$

The first and second terms in the objective function represent the contribution of exceptional elements and voids, respectively. Constraints 5.11 ensure that each part is assigned to a cell. Similarly, constraints 5.12 guarantee that each machine is allocated to a cell. Binary restrictions on the variables are imposed by constraints 5.13. The value of C is an overestimate of the number of cells. Since no arbitrary upper limit constraints are imposed on the number of parts or machines assigned in a cell, the model will identify the optimal number of cells and uncover natural groupings which exist in the data. Note that the first term in the objective function can also be stated as

$$w \sum_{c=1}^{C} \sum_{p=1}^{P} \sum_{m=1}^{M} a_{pm} Y_{mc} (1 - X_{pc})$$

i.e. the variables within and outside the brackets can be interchanged to compute the objective value. This will be used while decomposing the model in order to maintain consistency.

Solution methodology

If the part–machine matrix is small, the above model can be optimally solved by linearizing the terms in the objective function. For the efficient solution of larger problems (matrices of size, say, 400 × 200), Adil, Rajamani and Strong (1993a) provided a solution scheme called the assignment allocation algorithm. The solution to the above model is equivalent to block diagonalization minimizing the objective considered. Each block $c(c = 1,2, \ldots C)$ represents a cell. The variables Y_{mc} take a value of 1 if machine m is assigned to cell c or 0 otherwise. Similarly X_{pc} is 1 if part p is allocated to cell c or 0 otherwise. For a given assignment of machines and allocation of parts the objective function captures the contribution of the weighted sum of voids and exceptional elements. The nonlinearity of the terms in the objective function arises due to the product of these two decision variables, namely Y_{mc} and $X_{pc.}$ If one set of variables is known, say, Y_{mc}'s, the model can be solved for X_{pc} by simple inspection. Then by using the values of X_{pc} thus obtained, the model can be solved to obtain new values for the Y_{mc} variables. This procedure continues until convergence. Kasilingam (1989) proposed a similar approach for part–machine groupings by maximizing the compatibility indices between parts and machines. Srinivasan and Narendran (1991) improved the algorithm based on the assignment model presented in section 5.2 with a similar procedure. The algorithm proposed by Adil, Rajamani and Strong (1993a) is given below.

Algorithm

Step 1. As a starting solution, randomly assign the machines to the C cells which can be formed. If $C>M$, simply assign each machine to a separate cell. Thus, based on the assignment, the Y variables are known as, say, $\bar{Y}_{mc}$. For the given assignment compute the coefficient of the variables X_{pc} as follows:

$$B_{pmc} = w \cdot a_{pm}(1-\bar{Y}_{mc}) + (1-w)(1-a_{pm})\,\bar{Y}_{mc}$$

Step 2. (Allocation model). Solve the following model to obtain the optimal allocation of parts for a given machine assignment:

Min

$$\sum_{c=1}^{C} \sum_{p=1}^{P} \sum_{m=1}^{M} B_{pmc},\ X_{pc}$$

subject to:

$$\sum_{c=1}^{C} X_{pc} = 1, \quad \forall p \tag{5.14}$$

The above model is separable by parts and can be solved optimally simply by inspection. This can be interpreted as follows. For the current assignment of machines to cells, select a part and compute the number of voids and exceptional elements it will result in by assigning it to each of the cells. Denote the number of exceptional elements and voids as e_c and v_c, respectively, for any cell c. Compute the weighted objective value $(w \cdot e_c + (1-w)\,v_c)$ for all c. Assign the part to the cell which contributes to the minimum value. Once this allocation is performed for all parts the X variables are known as, say, $\bar{X}_{pc}$.

Step 3. (Assignment model). Solve the following model to obtain the optimal assignment of machines to cells for the allocation of parts determined in step 2.

Min

$$w \cdot \sum_{p=1}^{P} \sum_{m=1}^{M} a_{pm} + \sum_{c=1}^{C} \sum_{p=1}^{P} \sum_{m=1}^{M} D_{pmc}\, Y_{mc}$$

where

$$D_{pmc} = -wa_{pm}\bar{X}_{pc} + (1-w)(1-a_{pm})\bar{X}_{pc}$$

or interchanging the variables inside and outside brackets in the first term of the objective function gives

Min

$$\sum_{c=1}^{C}\sum_{p=1}^{P}\sum_{m=1}^{M} I_{pmc} Y_{mc}$$

where

$$I_{pmc} = wa_{pm}(1-\bar{X}_{pc}) + (1-w)(1-a_{pm})\bar{X}_{pc}$$

subject to:

$$\sum_{c=1}^{C} Y_{mc} = 1, \quad \forall m \tag{5.15}$$

The above model is separable by machines and can also be solved by inspection. The procedure outlined in step 2 can be used here in a similar way. At this step, assign each machine to the cell where it contributes to a minimum weighted objective value.

Step 4. If the objective value and solution do not change for the last two iterations, stop; else proceed to step 2.

Example 5.4

Consider the matrix in Fig. 3.1.

Steps 1 and 2. (Iteration 1). Let $C=5$ and $w=0.5$. As a starting solution, assign each machine to one cell leaving the last cell empty (Table 5.5).

Step 3. For the part allocation specified the optimal machine assignment is: machines 1,4 in cell 1, machine 2 in cell 2 and machine 3 in cell 3, and the remaining two cells are empty (Table 5.7).

Step 2. (Iteration 2). Now allocate parts for the new machine assignment obtained in step 3 (Table 5.8). For each part the number of exceptional elements and voids created by assigning it to a cell for the given machine assignment is shown in Table 5.6. The part is assigned to the cell which contributes to the minimum objective value. The allocation selected in the above case identifies parts 1 and 3 in cell 1, parts 2 and 5

Table 5.5 Starting solution to machine assignment

Cell number	$c=1$	$c=2$	$c=3$	$c=4$	$c=5$
Machines assigned	$m=1$	$m=2$	$m=3$	$m=4$	Empty

Table 5.6 Part allocation

Parts	*Exc*	*Void*	*Exc*	*Void*	*Exc*	*Void*	*Exc*	*Void*	*Exc*	*Void*
$p=1$	1	0	2	1	1	0	2	1	2	0
Obj	**0.5**		1.5		0.5		1.5		1	
$p=2$	1	1	0	0	1	1	1	1	1	0
Obj	1		**0**		1		1		0.5	
$p=3$	2	0	2	0	3	1	2	0	3	0
Obj	**1**		1		2		1		1.5	
$p=4$	1	1	1	1	0	0	1	1	1	0
Obj	1		1		**0**		1		0.5	
$p=5$	2	1	1	0	2	1	1	0	2	0
Obj	1.5		**0.5**		1.5		0.5		1	
Cell number	$c=1$		$c=2$		$c=3$		$c=4$		$c=5$	
Parts allocated	$p=1,3$		$p=2,5$		$p=4$		*Empty*		*Empty*	

Table 5.7 New machine assignment

Machines	*Exc*	*Void*	*Exc*	*Void*	*Exc*	*Void*	*Exc*	*Void*	*Exc*	*Void*
$m=1$	0	0	2	2	2	1	2	0	2	0
Obj	**0**		2		1.5		1		1	
$m=2$	2	1	1	0	3	1	3	0	3	0
Obj	1.5		**0.5**		2		1.5		1.5	
$m=3$	1	1	2	2	1	0	2	0	2	0
Obj	1		2		**0.5**		**1**		**1**	
$m=4$	1	1	1	1	2	1	2	0	2	0
Obj	**1**		1		1.5		1		1	
Cell number	$c=1$		$c=2$		$c=3$		$c=4$		$c=5$	
Parts allocated	$m=1,4$		$m=2$		$m=3$		*Empty*		*Empty*	

Table 5.8 Reallocation of parts

Parts	*Exc*	*Void*	*Exc*	*Void*	*Exc*	*Void*	*Exc*	*Void*	*Exc*	*Void*
$p=1$	1	1	2	1	1	0	2	0	2	0
Obj	**1**		1.5		0.5		1		1	
$p=2$	1	2	0	0	1	1	1	0	1	0
Obj	1.5		**0**		1		0.5		0.5	
$p=3$	1	0	2	0	3	1	3	0	3	0
Obj	**0.5**		1		2		1.5		1.5	
$p=4$	1	2	2	1	0	0	1	0	1	0
Obj	1.5		1.5		**0**		0.5		0.5	
$p=5$	1	1	1	0	2	1	2	0	2	0
Obj	1		**0.5**		1.5		1		1	
Cell number	$c=1$		$c=2$		$c=3$		$c=4$		$c=5$	
Parts allocated	$p=3$		$p=2,5$		$p=1,4$		*Empty*		*Empty*	

Table 5.9 Machine assignment from iteration 2

Machines	*Exc*	*Void*	*Exc*	*Void*	*Exc*	*Void*	*Exc*	*Void*	*Exc*	*Void*
$m=1$	1	0	2	2	1	1	2	0	2	0
Obj	**0.5**		2		1		1		1	
$m=2$	2	0	1	0	3	2	3	0	3	0
Obj	1		**0.5**		2.5		1.5		1.5	
$m=3$	2	1	2	2	0	0	2	0	2	0
Obj	1.5		2		**0**		1		1	
$m=4$	1	0	1	1	2	2	2	0	2	0
Obj	**0.5**		1		2		1		1	
Cell number	$c=1$		$c=2$		$c=3$		$c=4$		$c=5$	
Parts allocated	$m=1,4$		$m=2$		$m=3$		*Empty*		*Empty*	

in cell 2 and part 4 in cell 3 (the allocation selected is shown in bold). The two remaining cells are empty. Whenever a tie is encountered, the first minimum value is selected.
Step 3. For the part allocation obtained in the previous step, now assign machines (Table 5.9). The machine assignment obtained is the same as at the beginning of iteration 2. Thus the part allocation is the same and the procedure has converged.

In the matrix form the solution is given in Fig. 5.9. The above partition led to identification of three cells with three exceptional elements and an objective value of 1.5. An alternative solution is shown Fig. 3.3. This solution leads to forming two cells with two voids and one exceptional element. The objective value is again 1.5. Since equal weight has been given to an exceptional element and void, both are optimal solutions, but one might prefer to minimize the exceptional elements in comparison to voids. This can be accomplished by increasing the weight on exceptional elements to 0.7 and decreasing the weight on voids. In this case the solution shown in Fig. 3.3 will be obtained. Increasing the weight on exceptional elements leads to identification of large, loose cells, while decreasing the weight will identify small, tight cells.

By changing the value of w the designer can generate alternative solutions in a structured manner. A number of problems have been solved using this approach and a good partition is obtained for a value of $w=0.7$ in most cases. However, due to the nature of input data superior results may be obtained in the range 0.5–0.7 for some problems. A comparison of the results with other well known algorithms is provided in section 5.7. Also, it is possible to give different weights to different part/machine combinations to reflect the scenario when opportunity costs on machines (voids) and transportation costs of parts are not the same. This can be accomplished by replacing w in the above model with w_{pm}, where w_{pm} is the fraction representing the weight on an exceptional element corresponding to part p and machine m.

		Part (p)				
		3	2	5	1	4
Machine (m)	1	1			1	
	4	1		1		
	2	1	1	1		
	3				1	1

Fig. 5.9 Rearranged part–machine matrix.

5.6 EXTENDED NONLINEAR MODEL

Most part–machine matrices in real life are not perfectly groupable. This leads to the existence of bottleneck machines and exceptional parts. Since the objective of cell formation is to form mutually exclusive cells, these exceptional elements can be eliminated by selecting alternative process plans for parts, duplicating bottleneck machines in cells, part design changes or subcontracting the exceptional parts. The impact of alternative process plans and duplication of bottleneck machines is discussed next. Consider Fig. 3.3 which contains both an exceptional part (part 3) and a bottleneck machine (machine 1). If there were two copies of machine 1 available, the additional one could be assigned to the cell containing machines 2 and 4, thus completing part 3 within the cell. The procedures discussed so far have lumped all copies of a machine type as only one and were unable to consider this aspect. The new, rearranged partition is shown in Fig. 5.10.

If an additional copy of the machine is not available, one could consider identifying alternative process plans for the exceptional parts. For example, if there was an additional plan for part 3 where it required only machines 2 and 4, selecting this plan would have made it possible to process the part fully within the cell. Thus it is obvious that grouping of parts considering alternative process plans and also the available copies of machines enhances the possibility of identifying mutually independent cells.

The nonlinear model proposed can be extended to consider alternative process plans for parts and available copies of machines. Since we are considering reorganizing existing manufacturing activities, in the procedures developed so far we assume sufficient capacity is available and we are primarily interested in the minimum interaction between cells and the maximum number of machines visited by parts within each cell. This is achieved by minimizing the weighted sum of voids and exceptional elements. The extended model is given below.

	Part 1	4	3	5	2
Machine 1	1	1			
3	1	1			
1			1	0	0
2			1	1	1
4			1	1	0

Fig. 5.10 Rearranged partition with two copies of machine 1.

Simultaneous grouping model

Minimize

$$w \cdot \sum_{c=1}^{C} \sum_{p=1}^{P} \sum_{m=1}^{M} \sum_{r=1}^{R_p} a_{pm}^{r} X_{pc}^{r} (1 - Y_{mc}) +$$

$$(1 - w) \sum_{c=1}^{C} \sum_{p=1}^{P} \sum_{m=1}^{M} \sum_{r=1}^{R_p} (1 - a_{pm}^{r}) X_{pc}^{r} Y_{mc}$$

subject to :

$$\sum_{c=1}^{C} \sum_{r=1}^{R_p} X_{pc}^{r} = 1, \quad \forall p \tag{5.16}$$

$$\sum_{c=1}^{C} Y_{mc} \leqslant N_m, \quad \forall m \tag{5.17}$$

$$X_{pc}^{r}, Y_{\mathrm{mc}} = 0/1, \quad \forall_{p,m,c,r} \tag{5.18}$$

where r is the index for process plans, R_p is the number of process plans available for part p and N_m is the number of copies of machine type m, and

$$a_{pm}^{r} = \begin{cases} 1 & \text{if part } p \text{ requires processing on machine } m \text{ in process plan } r \\ 0 & \text{otherwise} \end{cases}$$

$$X_{pc}^{r} = \begin{cases} 1 & \text{if part } p \text{ is allocated to cell } c \text{ and process plan } r \text{ is selected} \\ 0 & \text{otherwise} \end{cases}$$

$$Y_{mc} = \begin{cases} 1 & \text{if machine } m \text{ is assigned to cell } c \\ 0 & \text{otherwise} \end{cases}$$

Constraints 5.16 guarantee that each part is allocated to one of the cells and only one process plan is selected for the part. Constraints 5.17

ensure that the number of machines assigned to cells does not exceed the available number of copies of machines. This model can again be solved optimally by linearizing the terms in the objective function (Adil, Rajamani and Strong, 1993b). The iterative assignment allocation using algorithm alternative process plans and $N_m = 1$ was tested for example problems and compared with the optimal solution obtained using the linearized model. It was observed that the initial input matrix affects the quality of the solution. Therefore Adil, Rajamani and Strong (1993c) developed a procedure based on simulated annealing which is robust and does not depend on the initial input matrix and arbitrary machine assignment. This section presents the linearized model. The simulated annealing approach is illustrated for the nonlinear model in the next chapter.

Linearized simultaneous grouping model

Minimize

$$w \sum_{p=1}^{P} \sum_{m=1}^{M} \sum_{c=1}^{C} \sum_{r=1}^{R_p} \delta^{+}_{pmcr} + (1-w) \sum_{p=1}^{P} \sum_{m=1}^{M} \sum_{c=1}^{C} \sum_{r=1}^{R_p} \delta^{-}_{pmcr}$$

subject to constraints 5.16 to 5.18 and the following:

$$\delta^{+}_{pmcr} \geqslant X^{r}_{pc} - Y_{mc}, \quad \forall_{p,m,c,r}, a^{r}_{pm} = 1 \tag{5.19}$$

$$\delta^{-}_{pmcr} \geqslant X^{r}_{pc} + Y_{mc} - 1, \quad \forall_{p,m,c,r}, a^{r}_{pm} = 0 \tag{5.20}$$

$$\delta^{+}_{pmcr}, \delta^{-}_{pmcr} \geqslant 0, \quad \forall_{p,m,c,r} \tag{5.21}$$

where

$$\delta^{+}_{pmcr} = \begin{cases} 1 & \text{if part } p \text{ is assigned to cell } c \text{ and uses plan } r \text{ and requires} \\ & \text{an inter-cell move for machine } m \text{ (i.e. an exceptional element)} \\ 0 & \text{otherwise} \end{cases}$$

$$\delta^{-}_{pmcr} = \begin{cases} 1 & \text{if part } p \text{ is assigned to cell } c \text{ and uses plan } r \text{ and does not} \\ & \text{require machine } m \text{ in cell } c \text{ (i.e. a void)} \\ 0 & \text{otherwise} \end{cases}$$

If the index r is dropped in the above model, the linear version of the model discussed in the previous section is obtained.

Example 5.5

Consider the part–machine matrix of Fig. 5.11 with alternative plans for parts. This problem will be solved for three different weights: $w = 0.5$ (case 1), $w = 0.3$ (case 2), and $w = 0.7$ (case 3). The solution obtained for each case is as follows.

Case 1. Part family 1: $\{1(2), 3(2)\}$; part family 2: $\{2(2), 4(2), 5(2)\}$
Machine group 1: $\{2, 4\}$; machine group 2: $\{1, 3\}$
Objective value = 0.5; number of voids = 1; number of exceptional elements = 0

Case 2. Part family 1: $\{1(2), 3(2)\}$; part family 2: $\{2(2), 4(2), 5(3)\}$
Machine group 1: $\{2, 4\}$; machine group 2: $\{1, 3\}$
Objective value = 0.3; number of voids = 0; number of exceptional elements = 1

Case 3. Objective value = 0.3; solution same as case 1

Thus, it can be seen that the model is able to consider a trade-off between voids and exceptional elements. If the above problem was solved using the generalized p-median model (Kusiak, 1987), with the objective to maximize the similarity, these solutions would not be distinguished.

5.7 OTHER MANUFACTURING FEATURES

The primary objective of the cell formation algorithms is to minimize the number of exceptional elements and voids. Alternate process plans or duplication of machines (if additional copies are available) are selected to reduce the number of exceptional elements (inter-cell moves), but the actions taken to eliminate an exceptional element have an impact on the complete cell system. Also, the actual number of inter-cell transfers is not determined by the number of exceptional elements alone. This is because the part sequence has not been considered. Other manufacturing features such as production volumes and capacities of machines in a cell have also been ignored. The part–machine matrix can be

Part (process plan)

Machine	1 (1)	1 (2)	1 (3)	2 (1)	2 (2)	3 (1)	3 (2)	4 (1)	4 (2)	5 (1)	5 (2)	5 (3)
1			1		1	1		1	1		1	1
2		1	1	1			1					1
3	1			1	1				1	1		1
4	1	1				1	1	1		1		

Fig. 5.11 Part–machine matrix for Example 5.5.

modified to include this additional information. For example, the part sequence on machines can be represented by defining a_{pm} as

$$a_{pm} = \begin{cases} k, & \text{if part } p \text{ visits machine } m \text{ for the } k\text{th operation} \\ 0, & \text{otherwise} \end{cases}$$

The modified clustered matrix is given in Fig. 5.12. It illustrates an example where three cells are identified with five exceptional elements. A machine may also be used by a part two or more times, as illustrated for part 4, which requires machine 2 for its second and fourth operations. Consecutive operations on the same machine can be treated as the same operation.

To illustrate the impact of sequence, consider the exceptional element appearing at the intersection of part 3 and machine 4. This operation will require two inter-cell moves: part 3 will travel to the second cell for the second operation and return to the first cell for the third operation. However, the exceptional element at the intersection of part 3 and machine 7 will require only one inter-cell move because it is the last operation on the part.

Assuming the option of changing the process plan has already been considered, we will emphasize the aspect of machine duplication. If an additional copy of machine 4 was available it could be placed in the first cell to eliminate the single exceptional element due to operation 2 of part 3 (i.e. two inter-cell moves). One could also have placed machine 4 in cell 3 to eliminate the exceptional element due to part 7 (again, two inter-cell moves). If there was only one additional copy where should it be placed? The information on production volumes and the material handling cost will provide the answer. Assuming the same unit handling cost, the machine will be assigned to the cell which processes more parts. In doing this we have not considered whether there is sufficient capacity available on machine 4 which was assigned to cell 2.

Machine \ Part	1	2	3	4	5	6	7	8
1	2	1	3					
2	1	2	1	2,4	4			
3					1	1		
4			2	1	3	2		2
5				3	2			
6							2	1
7			4				1	3,5
8								4

Fig. 5.12 Modified clustered matrix.

Thus, there is always a possibility of assigning both copies of machine 4 to cell 2 depending on production volumes and material handling cost. Similarly, if an additional copy of machine 2 was available it could be assigned to cell 2, thus resulting in a decrease of two exceptional elements (i.e. five inter-cell moves). If no additional copies of machines were available, the above partition would result in a total of ten inter-cell moves for a unit demand of all part types. If, however, the management is willing to buy one additional machine, which one should it be? Machine 2 would eliminate two exceptional elements (i.e. five inter-cell moves) as opposed to one (i.e. two inter-cell moves) by machine 4. Still, a single machine 2 might be more expensive than two or even three copies of machine 4. Thus, depending on part volumes, additional investment on machines can be economically justified if it results in a substantial saving on inter-cell material handling cost and on budget.

Finally, the most important aspect to consider is the impact of cell size on the intra- and inter-cell material handling cost. A reduction in cell size (fewer machines) reduces the intra-cell handling cost. On the other hand, the parts have to visit more cells to complete the processing. This increases the inter-cell handling cost. By balancing the inter- and intra-cell handling costs, one should be able to determine the optimal number of cells and cell sizes. Adil, Rajamani and Strong (1994) developed a two-stage procedure to consider many of these features (except duplication cost). In stage 1, a nonlinear model is developed to minimize the total intra- and inter-cell handling costs. In the calculation of the material handling costs, the factors considered are production quantity, effect of cell size on intra-cell handling, sequence of operations and multiple non-consecutive visits to the same machine. In stage 2, an integer programming model is developed to improve further the solution obtained in stage 1. The model considers the option to reassign the operations which resulted in exceptional elements in stage 1 and the extra copies of machines available.

5.8 COMPARISON OF ALGORITHMS FOR PART–MACHINE GROUPING

Miltenburg and Zhang (1991) reported the performance of nine well known algorithms on problems from the literature as well as on randomly generated test problems. This section reproduces the results obtained for the problems from the literature and compares them with the results obtained using the AAA. The evaluation criteria considered here will only be the primary measure, i.e. grouping measure. For the results of secondary measures, refer to Miltenburg and Zhang (1991). The nine algorithms and the data sets considered are given in

Table 5.10 Algorithms for part/machine grouping

Algorithm code	*Algorithm name*	*Algorithm used for part/machine grouping*
A1	Rank order clustering	ROC/ROC
A2	Similarity coefficient	SLC/ROC
A3	Similarity coefficient	SLC/SLC
A4	Modified similarity coefficient	ALC/ROC
A5	Modified similarity coefficient	ALC/ALC
A6	Modified rank order clustering	MODROC
A7	Seed clustering	ISNC*
A8	Seed clustering	SC-seed*
A9	Bond energy	BEA

* Algorithms not discussed in this text

Tables 5.10 and 5.11. The comparison of the primary measure is presented in Table 5.12. To test the performance of the AAA with data sets which range from well-structured to ill-structured, six 40×20 data sets (D1 to D6) (40 parts and 20 machines) were taken from Chandrasekaran and Rajagopalan (1989). The solutions obtained using the AAA were compared with the results obtained from two other algorithms, ZODIAC and GRAFICS (not discussed in this text) for the grouping efficiency and efficacy. The results are summarized in Table 5.13. Further, to test the computational efficiency of AAA, the six problems of varying structure were multiplied by 10 to get 400×200 matrices. All these problems were solved in less than 1 min. The number of iterations and computational times along with grouping measure values are shown in Table 5.14. The AAA shows favorably for the problems and performance measures compared as it is simple and less computer intensive.

Limitations of the AAA

The AAA is sensitive to the value of C and the initial input matrix. To see the effect of different starting solutions, the large problems L1 to L6 were solved for randomly generated starting solutions (for $C = M + 1$). The algorithm converged within six iterations for all problems. Although the algorithm is sensitive to the initial solution it yielded good solutions based on the grouping measure values obtained (Adil, Rajamani and Strong, 1993a). However, when C was varied, it greatly affected the quality of the solution. Most of the solutions obtained for different C were local optimum. A simulated annealing algorithm is proposed in the next chapter which is more robust and provides more consistent results. However, small problems can be solved optimally by linearizing the terms in the objective function.

Table 5.11 Well known problems from the literature

Problem code	*Reference*	*Number of parts (P)*	*Number of machines (M)*	*Density (d/PM)*
P1	Burbidge (1975)	43	16	0.18
P2	Carrie (1973)	35	20	0.19
P3	Black and Dekker (from Burbidge, 1975)	50	28	0.18
P4	Chandrasekaran and Rajagopalan (1986a)	20	8	0.38
P5	Chandrasekaran and Rajagopalan (1986b)	20	8	0.38
P6	Chan and Milner (1982)	15	10	0.31
P7	Ham, Hitomi and Yoshida (1985)	8	10	0.32
P8	Seifoddini and Wolfe (1986)	12	8	0.36

Table 5.12 Comparison of grouping measure

Problem	*Algorithms*									
	A1	A2	A3	A4	A5	A6	A7	A8	A9	AAA
P1	0.238	0.405	0.353	0.349	–	0.371	0.444	0.394	0.454	0.478 (0.7)*
P2	0.526	0.764	0.764	0.764	–	0.764	0.725	0.764	0.764	0.764 (0.7)
P3	0.176	0.215	0.176	0.183	–	0.176	0.239	0.198	0.250	–
P4	0.656	0.852	0.852	0.852	0.852	0.820	0.852	0.852	0.852	0.852 (0.7)
P5	0.569	0.569	0.569	0.569	0.569	0.569	0.569	0.569	0.569	0.569 (0.7)
P6	0.920	0.920	0.920	0.920	0.920	0.920	0.920	0.920	0.920	0.920 (0.7)
P7	0.812	0.812	0.812	0.812	0.812	0.812	0.812	0.812	0.812	0.812 (0.7)
P8	0.676	0.571	0.629	0.585	0.565	0.585	0.676	0.577	0.642	0.681 (0.5)

* Value of w used to solve the problem
[+] modified data could not be matched

5.9 RELATED DEVELOPMENTS

Mathematical models have received considerable attention in the last decade. The basic objective of these models is to maximize similarity/compatibility or minimize exceptional elements. As part of

Table 5.13 Comparison of grouping efficiency and grouping efficacy

Problem	*Grouping efficiency*			*Grouping efficacy*			*AAA*	
	ZODIAC	*GRAFICS*	*AAA*	*ZODIAC*	*GRAFICS*	*AAA*	*e*	*v*
D1	1	1	1	1	1	1	0	0
D2	0.952	0.952	0.952	0.851	0.851	0.851	10	11
D3/D4	0.9116	0.9116	0.9182	0.3785	0.7351	0.7297	23	17
D5	0.7731	0.7886	0.8753	0.2042	0.4327	0.5067	56	17
D6	0.7243	0.7914	0.8605	0.1823	0.4451	0.4459	65	17
D7	0.6933	0.7913	0.9085	0.1761	0.4167	0.4379	70	7

Table 5.14 Results for large problems

Problem	*CPU time (s)*	*Number of iterations*	*Number of cells*	*Measure* η_σ
L1	30.1	3	7	1.000
L2	33.3	3	7	0.839
L3/L4	33.7	3	8	0.688
L5	33.5	3	11	0.388
L6	54.1	5	12	0.299
L7	33.0	3	14	0.357

the input, information is required on the maximum number of machines and/or parts in each cell (Boctor, 1991; Kasilingam, 1989; Ribeiro and Pradin, 1993; Wei and Gaither, 1990). Some of these models consider assigning more than one copy of each machine type to cells (Kasilingam, 1989; Ribeiro and Pradin, 1993; Wei and Gaither, 1990). The basic assumption in all the procedures discussed in Chapters 3 to 5 was that there was sufficient capacity available in each cell to process all the parts, and when more than one copy was available, the additional copies were assigned to cells such that the exceptional elements were minimized. The machine requirements for parts in each cell were not computed to identify a cost-effective assignment. Relatively few models consider capacity restrictions at this stage (Wei and Gaither, 1990).

The cell design process is relatively complex and often proceeds in stages. As stated earlier, the algorithms for cell formation provide the first rough-cut groups. The exceptional elements and each group can be individually considered in a more detailed analysis that includes other manufacturing aspects such as part sequence, processing times, machine capacities and the trade-off between the purchase of additional machines and material handling to make groups independent. A few procedures which work on further improving the solution obtained by cell formation algorithms are by Sule (1991), Kern and Wei (1991), Logendran (1992), Shafer, Kern and Wei (1992). These procedures assume that the option to change the parts processing plans to suit the cell has already been considered. The importance of considering alternative process plans during cell formation has been addressed by only a few researchers (Kusiak, 1987; Kasilingam, 1989; Kusiak and Cho, 1992; Adil, Rajamani and Strong, 1993b; Adil, Rajamani and Strong, 1993c).

5.10 SUMMARY

A mathematical programming statement of a seemingly small cell formation problem becomes large, combinatorial and NP-complete, and hence most of the procedures developed in Chapters 2 to 4 are heuristics. These heuristics suffer from one or more drawbacks. For example, the matrix manipulation algorithms (Chapter 3) require manual intervention to identify part families and machine groups. This becomes difficult for large matrices that are not perfectly groupable. The clustering methods (Chapter 4) require a large amount of data storage and computation of similarity matrices, and do not identify part families and machine groups simultaneously. Also, they suffer from the chaining problem. This chapter introduced a few mathematical models for optimally identifying part families and/or machine groups. The p-median, assignment and quadratic models adopt a sequential approach by identifying part families (or machine groups) first, followed by some procedure for identifying the machine groups (or part families). The objective of all these models is to maximize similarity, but they differ in considering the interaction between parts (or machines) within a family (or group). Graph-based methods were also briefly introduced.

These models and the heuristic procedures do not necessarily consider the objectives of cell formation explicitly. For this purpose, a nonlinear model was proposed which overcomes most of the drawbacks of the algorithms proposed in Chapters 3 to 5. The objective of this model is to minimize explicitly the weighted sum of voids and exceptional elements. By changing weights for voids and exceptional elements the user has the flexibility to form large, loose cells or small, tight cells to suit the situation. This model identifies part families and machine groups simultaneously without any manual intervention. The model also identifies parts and machines which are not suitable for assigning to cells. An efficient iterative algorithm (the AAA) was presented for partitioning large matrices. The results obtained using the AAA compare favorably with well-known algorithms in the literature. The AAA is simple and less computer intensive. The nonlinear model

	Part							
Machine	1	2	3	4	5	6	7	8
1					1	1		1
2	1			1	1			1
3	1	1	1				1	
4	1		1	1				
5	1	1	1			1		

Fig. 5.13 Part–machine matrix for Q5.1.

was further extended to consider alternative process plans and additional copies of machines during the cell formation process. The impact and importance of considering the other manufacturing features was briefly addressed.

PROBLEMS

5.1 Consider the part–machine matrix in Fig. 5.13. Use the p-median model to identify two machine groups. Use the assignment model to identify part families and machine groups. What advantage or disadvantage does the assignment model have over the p-median model?

5.2 Apply the quadratic programming model for the data in Q 5.1. Identify the corresponding part families.

5.3 Compare and contrast the nature of part families and machine groups obtained using: single linkage clustering and the linear clustering algorithm; average linkage clustering and the quadratic programming model; the assignment model, p-median model and the quadratic model.

5.4 Represent the data provided in Q 5.1 as a bipartite graph. Write the corresponding quadratic model to identify part families and machine groups.

5.5 Explain the importance of considering voids and exceptional elements explicitly in the process of manipulating the matrix instead of the similarity measure.

5.6 Apply the AAA to the data in Q 5.1 to obtain the groupings. Compare this solution with the optimal solution obtained using the linearized model for $w = 0.3$ and $w = 0.7$. Comment on the nature of groupings obtained.

5.7 Consider the part–machine matrix of Fig. 5.14, where five parts have two or three alternative process plans. Extend the concept of AAA to consider alternative process plans. Do you foresee any problem with this procedure when alternative plans are considered? Compare the above solution with the optimal solution obtained using the linearized model.

Part (process plan)

Machine	1 (1)	1 (2)	1 (3)	2 (1)	2 (2)	3 (1)	3 (2)	4 (1)	4 (2)	5 (1)	5 (2)
1		1	1	1		1	1	1			
2	1	1			1			1			1
3	1		1				1	1	1	1	
4				1	1	1			1	1	1

Fig. 5.14 Part–machine matrix for Q5.7.

Part

Machine	1	2	3	4	5	6	7	8
1	2				1	3		
2			1				1	
3				2				1
4		3					2	
5								4
6	1	2,4			2	1	4	
7		1	2			2	3	2
8				1		4		3,5

Fig. 5.15 Part–machine matrix for Q5.8.

5.8 Consider the part–machine matrix of Fig. 5.15 with the sequence of visits shown. Develop a mathematical model for machine grouping to minimize the sum of intra- and inter-cell moves considering the sequence of machine visits. What solution procedure do you suggest to solve the model proposed? State the assumptions made for developing the model.

REFERENCES

Adil, G.K., Rajamani, D. and Strong, D. (1993a) AAA–an assignment allocation algorithm for cell formation. Univ. Manitoba, Canada. Working paper.

Adil, G.K., Rajamani, D. and Strong, D. (1993b) An algorithm for cell formation considering alternate process plans, in *Proceedings of IASTED International Conference*, Pittsburgh, PA, PP. 285–8.

Adil, G.K., Rajamani, D. and Strong, D. (1993c) Cell formation considering alternate routings. Univ. Manitoba, Canada. Working paper.

Adil, G.K., Rajamani, D. and Strong, D. (1994) A two stage approach for cell formation considering material handling. Univ. Manitoba, Canada. Working paper.

Boctor, F.F. (1991) A linear formulation of the machine part cell formation. *International Journal of Production Research*, **29**(2), 343–56.

Burbidge, J.L., (1975) *The Introduction of Group Technology*, Wiley, London.

Burbidge, J.L. (1993) Comments on clustering methods for finding GT groups and families. *Journal of Manufacturing Systems*, **12**(5), 428–9.

Carrie, A.S. (1973) Numerical taxonomy applied to group technology and plant layout. *International Journal of Production Research*, **11**(4), 399–416.

Chan, H.M. and Milner, D.A. (1982) Direct clustering algorithm for group formation in cellular manufacture. *Journal of Manufacturing Systems*, **1**(1), 65–75.

Chandrasekaran, M.P. and Rajagopalan, R. (1986a) MODROC: an extension of rank order clustering algorithm for group technology. *International Journal of Production Research*, **24**(5), 1221–33.

Chandrasekaran, M.P. and Rajagopalan, R. (1986b) An ideal seed non-hierarchical clustering algorithm for cellular manufacturing. *International Journal of Production Research*, **24**(2), 451–64.

Chandrasekaran, M.P. and Rajagopalan, R. (1989) Groupability: an analysis of the properties of binary data matrices for group technology. *International Journal of Production Research*, **27**(7), 1035–52.

Ham, I., Hitomi, K. and Yoshida, T. (1985) *Group Technology: Applications to Production Management*, Kluwer-Nijhoff Publishing, Boston.

Kasilingam, R.G. (1989) Mathematical programming approach to cell formation problems in flexible manufacturing systems. Univ. Windsor, Canada. Doctoral dissertation.

Kern, G.M. and Wei, J.C. (1991) The cost of eliminating exceptional elements in group technology cell formation. *International Journal of Production Research*, **29**(8), 1535–47.

Kumar, K.R., Kusiak, A. and Vannelli, A. (1986) Grouping of parts and components in flexible manufacturing systems. *European Journal of Operational Research*, **24**, 387–97.

Kusiak, A. (1987) The generalized group technology concept. *International Journal of Production Research*, **25**(4), 561–9.

Kusiak, A. and Cho, M. (1992) Similarity coefficient algorithms for solving the group technology problem. *International Journal of Production Research*, **30**(11), 2633–46.

Kusiak, A. and Chow, W.S. (1988) Decomposition of manufacturing systems. *IEEE Journal of Robotics and Automation*, **4**(5), 457–71.

Kusiak, A., Vannelli, A. and Kumar, K.R. (1986) Clustering analysis: models and algorithms. *Control and Cybernetics*, **15**(2), 139–54.

Lee, J.L., Vogt, W.G. and Mickle, M.H. (1982) Calculation of shortest paths by optimal decomposition. *IEEE Transactions on Systems, Man and Cybernetics*, **12**, 410–15.

Logendran, R. (1992) A model for duplicating bottleneck machines in the presence of budgetary limitations in cellular manufacturing. *International Journal of Production Research*, **30**(3), 683–94.

Miltenburg, J. and Zhang, W. (1991) A comparative evaluation of nine well known algorithms for solving cell formation problem in group technology. *Journal of Operations Management*, **10**(1), 44–72.

Rajagopalan, R. and Batra, J.L. (1975) Design of cellular production systems: a graph theoretic approach. *International Journal of Production Research*, **13**(6), 567–79.

Ribeiro, J.F.F. and Pradin, B. (1993) A methodology for cellular manufacturing design. *International Journal of Production Research*, **31**(1), 235–50.

Seiffodini, H. and Wolfe, P.M. (1986) Application of the similarity coefficient method in group technology. *IIE Transactions*, **18**, 271–7.

Shafer, S.M., Kern, G.M. and Wei, J.C. (1992) A mathematical programming approach for dealing with exceptional elements in cellular manufacturing. *International Journal of Production Research*, **30**(5), 1029–36.

Song, S. and Hitomi, K. (1992) GT cell formation for minimizing the intercell part flow. *International Journal of Production Research*, **30**(12), 2737–53.

Srinivasan, G. and Narendran T.T. (1991) GRAFICS–a nonhierarchical clustering algorithm for group technology. *International Journal of Production Research*, **29**(3), 463–78.

Srinivasan, G., Narendran, T.T and Mahadevan, B. (1990) An assignment model for the part families problem in group technology. *International Journal of Production Research*, **28**(1), 145–52.

Sule, D.R. (1991) Machine capacity planning in group technology. *International Journal of Production Research*, **29**(9), 1909–22.

Vannelli, A. and Kumar, K.R. (1986) A method for finding minimal bottleneck cells for grouping part–machine families. *International Journal of Production Research*, **24**(2), 387–400.

Vohra, T., Chen, D.S., Chang, J.C. and Chen, H.C. (1990) A network approach to cell formation in cellular manufacturing. *International Journal of Production Research*, **28**(11), 2075–84.

Wei, J.C. and Gaither, N. (1990) An optimal model for cell formation decisions. *Decision Sciences*, **21**(2), 416–33.

Wu, N. and Salvendy, G. (1993) A modified network approach for the design of cellular manufacturing systems. *International Journal of Production Research*, **31**(6), 1409–21.

CHAPTER SIX

Novel methods for cell formation

The cell formation problem is a combinatorial optimization problem. The optimization algorithms yield a globally optimal solution in a possibly prohibitive computation time. Hence, a number of heuristics were proposed in earlier chapters. The heuristics presented are all tailored algorithms capturing expert skill and knowledge to the specific problem of identifying part families and machine groups. These heuristics yield an approximate solution in an acceptable computation time. However, these algorithms are sensitive to the initial solution, the groupability of the input part–machine matrix and the number of cells specified. Thus there is usually no guarantee that the solution found by these algorithms is optimal. The key to dealing with such problems is to go a step beyond the direct application of the expert skill and knowledge and make recourse to special procedures which monitors and directs the use of this skill and knowledge. Five such procedures have recently emerged: simulated annealing (SA), genetic algorithms (GA), neural networks (NN), tabu search and target analysis.

Simulated annealing derives from physical science; genetic algorithms and neural networks are inspired by principles derived from biological sciences; tabu search and target analysis stem from the general tenets of intelligent problem solving (Glover and Greenberg, 1989). These procedures are random search algorithms and are applicable to a wide variety of combinatorial optimization problems. This chapter introduces SA, GA and NN in the context of cell formation. These algorithms incorporate a number of aspects related to iterative algorithms such as the AAA. However, the main difference is that these random search algorithms provide solutions which do not depend on the initial solution and have an objective value closer to the global optimum value. It is important to recognize that a randomized search does not necessarily imply a directionless search. The nonlinear mathematical model presented in Chapter 5 is the problem for which these procedures are implemented.

6.1 SIMULATED ANNEALING

The simulated annealing approach is based on a Monte Carlo model used to study the relationship between atomic structure, entropy and temperature during the annealing of a sample of material. The physical process of annealing aims at reducing the temperature of a material to its minimum energy state, called 'thermal equilibrium'. The annealing process begins with a material in a melted state and then gradually lowers its temperature. At each temperature the solid is allowed to reach thermal equilibrium. The temperature must not be lowered too rapidly, particularly in the early stages, otherwise certain defects can be frozen in the material and the minimum energy state will not be reached. The lowering of the temperature is analogous to decreasing the objective value (for a minimization problem) by a series of improving moves. To allow a temperature to move slowly through a particular region corresponds to permitting non-improving moves to be selected with a certain probability, a probability that diminishes as the objective value decreases.

The design of SA depends on three key concepts (Francis, McGinnis and White, 1992). The first is referred to as the 'temperature' and is essentially the parameter that controls the probability that a cost-increasing solution will be accepted (for a minimization problem). During the course of SA the temperature will be reduced periodically, reducing the probability of accepting a cost-increasing solution. The solution in this case refers to the part–machine groupings. The cost-increasing solution refers to the weighted sum of voids and exceptional elements. The second key concept is 'equilibrium', or a condition in which it is unlikely that further significant changes in the solution will occur with additional sampling. For example, if a large number of interchanges have been attempted at a given temperature without finding a better solution, it is unlikely that additional sampling will be productive. The third key concept is the 'annealing schedule', which defines the set of temperatures to be used and how many interchanges to consider (or accept) before reducing the temperature. If there are too few temperatures or not enough interchanges are attempted at each temperature, there is a great likelihood of stopping with a suboptimal solution. In the context of the cell formation problem, SA resembles the AAA, with one very important difference: in SA a solution which corresponds to an increase in cost or objective value is accepted in a limited way. Thus there is at least some chance that an unlucky choice of intermediate solution will not cause the search to be trapped at a suboptimal solution.

This section presents an implementation of the SA to obtain groupings of parts and machines (Adil, Rajamani and Strong, 1994). The main steps in this algorithm are: initial solution, generation of neighborhood solution, acceptance/rejection of generated solution, and termination.

Initial solution

The maximum number of cells to be formed C is first specified. An initial machine assignment is generated. Machines are assigned to the cells using a predefined rule. For example, initially each machine can be assigned to a separate cell or the machines could be assigned to cells randomly. For this machine assignment, an initial part allocation is obtained by solving the allocation subproblem (as in the AAA, Chapter 5). Thus, an initial solution (part families and machine groups) and the objective function value are obtained.

Generation of a neighborhood solution

At each subsequent iteration, one machine is moved from the current cell to another cell, forming a new machine assignment. The machine to be moved and the cell for this machine are selected randomly (Boctor, 1991). Parts are allocated for this new machine assignment and the objective value is computed.

Acceptance / rejection of the generated solution

The generated solution (new part families and machine groups) is accepted if the objective function value improves. If the objective function value does not improve, the solution is accepted with a probability depending on the temperature, which is set to allow the acceptance of a large proportion of generated solutions at the beginning. Then, the temperature is modified to reduce the probability of acceptance. At each cooling temerature many moves are attempted and the algorithm stops when predefined conditions are met.

Termination

If the specified maximum iterations are reached or the acceptance ratio (defined below) is below a predetermined value, the algorithm is stopped.

Selection of simulated annealing parameters

The implementation of the SA algorithm requires the following parameters to be specified (Laarhoven and Aarts, 1987). The choice of these parameters is referred to as a 'cooling schedule'. In this implementation the cooling schedule is defined in the following way (Adil, Rajamani and Strong, 1994).

Initial temperature T_o

The initial temperature T_o is taken in such a way that virtually all transitions are accepted. An acceptance ratio R is defined as the number of accepted transitions divided by the number of proposed transitions. The value of T_o is set in such a way that the initial acceptance ratio R_o is close to unity. Usually the value of T_o is of the order of the expected objective function value. The value of T_o is increased or decreased to bring the acceptance ratio for the first ten iterations to between 0.95 and 1.0.

Length of Markov chain L_i (at iteration i)

The length of Markov chains L_i are controlled in such a way that for each value of temerature T_i a minimum number of transitions should be accepted, i.e. L_i is determined such that the number of accepted transitions is at least AT_{min}. However, as T_i approaches 0, transitions are accepted with decreasing probability and thus one eventually obtains $L_i \rightarrow \infty$ for $T_i \downarrow$ (approaches) 0. Consequently, L_i is bounded by some constant L (usually a chosen polynomial in the problem size) to avoid extremely long Markov chains for low cooling temperatures. We define $L = aM^2$, where M is the total number of machines and a is a constant. The value of AT_{min} should be high enough to ensure that equilibrium is reached at each temperature. The higher the value chosen for AT_{min}, the better the expected quality of the solution.

Rule for changing the current value of temperature

To ensure slow cooling the temperature decrements should be gradual. A frequently used decrement rule is given by $T_i = \alpha \cdot T_{i-1}$, where α is a constant smaller than, but close to unity. Also, if faster cooling is desired, AT_{min} is given a high value and α is given a lower value. Thus, for fast cooling the Markov chains at each temperature should be longer.

Termination

Defining the value of the final temperature is the stopping criterion used most often in SA. In this implementation, the final temperature is not chosen a priori. Instead, the annealing is allowed to continue until the system is frozen by one of the following criteria:

- the maximum number of iterations (temperature) i_{max};
- the acceptance ratio is smaller than a given value R_f at a given temperature; that is,
- the objective of the last accepted temperature transition is identical for a number of iterations (kept at 20 iterations).

Detailed steps for the implementation of SA are presented below.

Algorithm

Step 1. Initialization. Set the annealing parameters and generate an initial solution.

(a) define the annealing parameters T_o, AT_{min}, α, i_{max} and R_f.
(b) initialize iteration counter $i = 0$.
(c) generate an initial machine assignment and allocate parts by solving the allocation model (get SOL^0, OBJ^0).

Step 2. Annealing schedule. Execute outer loop, i.e. steps (a)–(g) below, until conditions in step 2(g) are met.

(a) Initialize inner loop counter $l = 0$ and the accepted number of transitions $AT = 0$.
(b) Initialize solution for inner loop, $SOL_0 = SOL^i$, $OBJ_0 = OBJ^i$.
(c) Equilibrium. Execute inner loop, i.e. steps (i)–(v) below, until conditions in step (v) are met.
    (i) Update $l = l + 1$.
    (ii) Generate a neighboring solution by perturbing the machine assignment and obtaining a parts allocation for the new machine assignment (get SOL_1, OBJ_1).
    (iii) Set $\delta = OBJ_l\text{-}OBJ_{l-1}$.
    (iv) If $\delta \leqslant 0$ or random $(0,1) \leqslant e^{-\delta/T_i}$ then
    SOL_1 and OBJ_1 are accepted
    update $AT = AT + 1$
    else
    solution is rejected, $SOL_l = SOL_{l-1}$, $OBJ_l = OBJ_{l-1}$.
    (v) If one of the following conditions holds true, $AT \geqslant AT_{min}$ or $L \geqslant M^2$ (M = number of machines), then terminate the loop and go to step 2(d), else continue the inner loop and go to 2(c)(i).
(d) Update $i = i + 1$.
(e) Update $SOL^i = SOL_l$, $OBJ^i = OBJ_l$.
(f) Reduce the cooling temperature $T_i = \alpha\, T_{i-1}$.
(g) If one of the following conditions holds true, $i \geqslant i_{max}$ or the acceptance ratio (defined as $AT/1) \leqslant R_f$ or the objective value for the last ten iterations remains the same, then terminate the outer loop and go to step 3, else continue the outer loop and go to step 2(a).

Step 3. Print the best solution obtained and terminate the procedure.

Example 6.1

Consider the part–machine matrix given in Fig. 6.1. Only the first iteration is illustrated.

		Part 1	2	3	4
Machine	1	1	1	0	0
	2	0	0	1	1
	3	0	0	1	0
	4	1	1	0	0

Fig. 6.1 Part–machine matrix for Example 6.1.

Step 1. Initialization

(a) Define the annealing parameters: $T_0 = 2$, $AT_{min} = 5$, $\alpha = 0.6$, $i_{max} = 50$ and $R_f = 0.01$.
(b) Initialize iteration counter $i = 0$.
(c) Let $C = 5$, assign each machine to a separate cell leaving the last cell empty. Allocate parts to cells as with the AAA. The initial solution (SOL^0) is:

cell 1: machine 1, parts 1 and 2
cell 2: machine 2, parts 3 and 4
cell 3: machine 3, no parts
cell 4: machine 4, no parts
cell 5: no machines, no parts.

The initial objective value (OBJ^0) (weighted sum of voids and exceptional elements for $w = 0.7$) is 2.1.

Step 2. Annealing schedule. Execute outer loop, i.e. steps 2(a)–(g), until conditions in step 2(g) are met.

(a) Initialize inner loop counter $l = 0$ and accepted number of transitions $AT = 0$.
(b) Initialize solution for inner loop, $SOL_0 = SOL^0$, $OBJ_0 = OBJ^0 = 2.1$.
(c) Equilibrium. Execute inner loop, i.e. steps 2(c)(i)–(v) until conditions in step 2(c)(v) are met.

(i) Update $l = 0 + 1 = 1$.
(ii) Generate a neighboring solution by moving machine 3 from cell 3 to cell 1. The new parts allocation is obtained (SOL_1). The objective value corresponding to this solution is $OBJ_1 = 2.1$.
(iii) Set $\delta = OBJ_1 - OBJ_0 = 2.1 - 2.1 = 0$.
(iv) Since $\delta \leqslant 0$, SOL_1 and OBJ_1 are accepted. Update $AT = 0 + 1 = 1$.
(v) Since $AT = 1$ is not $\geqslant AT_{min} = 5$, go to step 2(c)(i). Iterate steps 2(c)(i) to (c)(v) until $AT = 5$ (in this case). At the end of iteration 1; $T = 2$; acceptance ratio = 1.

(d) Update $i = 0 + 1 = 1$.
(e) Update $SOL^1 = SOL_5$, $OBJ^1 = OBJ_5$.

(f) Reduce the cooling temperature $T_1 = 0.60$. $T_0 = 0.6 \times 2 = 1.2$.
(g) Check the following conditions:
$i = 1$ is not $\geqslant i_{max} = 50$; the acceptance ratio ($AT/1$) is not $\leqslant R_f = 0.01$; the objective value of the last ten iterations does not remain the same, hence go to step 2(a).

(a) Step 2.1. Initialize inner loop counter $l = 0$ and accepted number of transitions $AT = 0$.
(b) Step 2.2. Initialize solution for inner loop, $SOL_0 = SOL^1$, $OBJ_0 = OBJ^1$.
(c) Step 2.3. Continue the process as before.

The algorithm terminates after six iterations (steps 2(a)–(g)), with an acceptance ratio = 0. Go to step 3.

Step 3. The $OBJ^5 = 0.3$ and the SOL^5 is:

cell 1: machines 1 and 4, parts 1 and 2
cell 5: machines 3 and 2, parts 3 and 4
cells 2, 3 and 4 are empty.

Only the required number of cells are identified. The solution identifies one void and no exceptional elements.

The above algorithm has been tested on data sets from section 5.8. The results obtained were consistent even for different values of C (Adil, Rajamani and Strong, 1994). This implementation has also been tested in the context of alternate process plans and it provided good results for all the problems tested. A more careful study of the influence of the parameters is warranted to provide the user with guidelines on the selection of parameters for different matrix sizes and types of data.

6.2 GENETIC ALGORITHMS

A genetic algorithm is a random search technique developed by Holland (1992). It is based on an interesting conclusion that where robust performance is desired (and where is it not?), nature does it better; the secrets of adaptation and survival are best learned from careful study of biological examples (Goldberg, 1989). Thus, it was originally inspired by an analogy with the process of natural evolution. In evolution, the problem that each species faces is one of searching for beneficial adaptations to a complicated and changing environment. The knowledge that each species has gained is embodied in the make-up of the chromosomes of its members. Genetic operations, viz. random mutation, inversion of chromosomal material and exchange of chromosomal material between two parent chromosomes, are used in the search for beneficial adaptation when the parents reproduce and evolution takes place. Likewise, a genetic algorithm combines the survival-of-the-fittest among solution structures with a structured, yet randomized, information exchange and creates offsprings, viz. new sets of solutions using bits and pieces of the fittest of the old solution structures. The offsprings displace weak solutions during each generation. Thus a genetic algorithm somewhat mimics at high speed the underlying genetic processes (Venugopal and Narendran, 1992).

The design of a GA depends on six key concepts: representation, initialization, evaluation function, reproduction, crossover and mutation (Gupta *et al.*, 1994). In a simple GA, a candidate solution is represented by a sequence of genes or binary numbers and is known as a chromosome. A chromosome's potential as a solution is determined by its fitness function which evaluates a chromosome with respect to the objective function of the optimization problem at hand. A judiciously selected set of chromosomes is called a 'population' and the population at a given time is a 'generation'. The population size remains fixed from generation to generation and has a significant impact on the performance of the GA. The GA operates on a generation and, generally, consists of three main operations: reproduction (selection of copies of chromosomes according to their fitness value); crossover (exchange of a portion of the chromosomes); and mutation (a random modification of the chromosome). The chromosomes resulting from these three operations, often known as offsprings or children, form the population of the next generation. The process is then iterated the desired number of times (usually up to the point where the system ceases to improve or the population has converged to a few well-performing sequences). Thus, it can be seen that the GA does not evaluate and improve a single solution, instead it analyses and modifies a population of solutions simultaneously. This ability of a GA to operate on many solutions simultaneously and to gather information from all points to direct the search enables the algorithm to escape from a local optimum. The different key concepts in the context of cell formation are discussed next. The implementation and concepts presented are abstracted from Gupta *et al.* (1994).

Representation

A problem can be solved once it can be represented in the form of a solution string. The bits in the chromosome could be binary, integers or a combination of characters. For the cell formation considered, each gene represents a cell number and the positioning of the gene in the chromosome represents the machine number or part number. The first M positions correspond to the machines and the last P positions correspond to the parts (M and P are the numbers of machines and parts). For example, when $M = 5$ and $P = 5$ the chromosome $(2, 1, 2, 1, 3, 2, 2, 3, 3, 1)$ represents a three-cell solution with the following machines and parts in each cell (Gupta and Rajamani, 1994):

cell = 1: machines 2, 4; part 5
cell = 2: machines 1, 3; parts 1, 2
cell = 3: machine 5; parts 3, 4

Initialization

The initialization process can be executed with either a randomly created population or a well adapted (seeded) population. In this implementation, an initial population of the desired size (PPSZ) is generated randomly.

Evaluation or fitness function

In a GA a fitness function value is computed for each string in the population and the objective is to find a string with the maximum value. The objective of the cell formation problem is the minimization of the weighted sum of voids and exceptional elements. It is necessary to map this objective to a fitness function through one or more mappings. The following transformation is applied here (Goldberg, 1989):

$$f(t) = \begin{cases} f_{max} - g(t), & \text{when } g(t) < f_{max} \\ 0, & \text{otherwise} \end{cases} \tag{6.1}$$

where $g(t)$ is the objective value (weighted sum of voids and exceptional elements) and f_{max} is the largest objective function value in the current generation.

Selection and reproduction

Strings with higher fitness values are selected for crossover and mutation using the selection process of stochastic sampling without replacement (Goldberg, 1989). Booker's (1987) investigation of the stochastic sampling without replacement method demonstrated its superiority over other selection schemes (deterministic sampling, stochastic sampling with replacement and stochastic tournament), and as a result this process is used here. In this process, the expected count of each string $e_i = (F_i/F)\,\text{PPSZ}$, where F_i is the fitness value of the i string, F is the average fitness value of all the strings in the population and PPSZ is the population size. Each string is then allocated samples according to the integer part of e_i and the fractional part of e_i is treated as the probability of an additional copy in the next generation. For example, if $e_1 = 2.25$, then the next generation will receive two copies of this string and has a probability of 0.25 of receiving the third.

Crossover and mutation

The chromosomes to be crossed and the crossing point(s) are selected randomly. The single-point crossover technique is used to illustrate the concepts. The crossover is done with a probability called the crossover probability (PCRS). For example, consider two parent chromosomes with a crossing point of 3:

parent 1: 1 2 3 | 2 2 1 2 2
parent 2: 2 3 1 | 1 1 1 3 3

The crossover operator generates two children:

child 1: 1 2 3 1 1 1 3 3
child 2: 2 3 1 2 2 1 2 2

To build in randomness, mutation is done with a low probability (PMUT1). Two random numbers r_1 and r_2 are generated such that $1 \leqslant r_1 \leqslant M+P$ (total number of parts plus machines) and $1 \leqslant r_2 \leqslant C$ (maximum number of cells specified). The cell number corresponding to the machine or part specified by r_1 is replaced with the cell number r_2.

Replacement strategy

After every crossover and mutation, new strings are created. Poor performing offsprings are replaced in the new generation with a replacement strategy. Several strategies have been suggested (Liepins and Hillard, 1989); the most common is probabilistically to replace the poorest performing sequences in the previous generation. A crowding strategy probabilistically replaces either the most similar parent or the most similar chromosome in the previous generation, whereas the elitist strategy appends the best performing chromosome of the previous generation to the current population and thereby ensures that the sequences with the best performance always survive intact into the next generation.

A combination of the above strategies is used in this implementation. By using crossover and mutation, a pool of offsprings is generated to create a new population. If all the offsprings outperform every existing chromosome in the old population, then all the offsprings replace the existing chromosomes in the new population. On the other hand, if only some of them fare better, then they replace an equal number of existing chromosomes. Usually, the strings with lowest performance are replaced, while the offspring is selected through a measure called the acceptance probability β. If S_j is the population with $f(S_j)$ (objective value of S_j) $>f(S_i)$ (objective value of S_i), then $\beta = \exp\{-f(S_j) - f(S_i)\}$. The replacement strategy thus establishes that only the best performing chromosomes of the previous generation are carried into the next generation.

Convergence policy (termination)

Clearly, as generations proceed the population is filled with better performing chromosomes. An ideal genetic algorithm should maintain a high degree of diversity within the population as it iterates from one generation to the next, otherwise the population may converge prematurely before the desired solution is found (Grefenstette, 1987). Baker (1987) observed that rapid convergence often occurs after a sequence in which a small group of individuals contributes a large number of offsprings to the next generation. In turn, when a large number of offsprings from one sequence deprives the other sequences from producing offsprings, it results in a rapid loss of diversity and premature convergence causing a problem called the genetic drift (Booker, 1987). Consequently, a number of proposals have been made to prevent premature convergence and improve the search capabilities of genetic algorithms. One proposal is to increase the population size, but there is a limit to increasing the population size in order to obtain efficient solutions (Booker, 1987). Grefenstette (1987) suggested an entropic measure H_i in a population of chromosomes, i.e. for each machine of part i, compute the entropic measure H_i in the current population as follows:

$$H_i = \frac{\sum_{c=1}^{C} (n_{ic}/S) \log (n_{ic}/S)}{\log C} \tag{6.2}$$

where n_{ic} is the number of chromosomes in which machine of part i is assigned to cell c in the current population, S is the population size, and C is the maximum number of cells. The divergence H is then calculated as

$$H = \frac{\sum_{i=1}^{M+P} H_i}{M+P} \tag{6.3}$$

As the population converges, $H \to 0$. So monitoring divergence after each generation can avoid premature convergence. If diversity falls below a predetermined value, say 0.005, mutation is performed with a high probability PMUT2, so as to maintain a diversity in the population.

Parameter values

The values of a variety of parameters and policies like crossover rate (PCRS), mutation rate (PMUT1, PMUT2), population size (PPSZ), number of generations (XGEN), replacement policy and divergence policy play a crucial role in the successful implementation of a genetic algorithm. The importance of selecting the appropriate values for these parameters was reported by De Jong (1975). The detailed steps of the implementation are presented below. This implementation is intended to introduce the reader to GAs and has not yet been well tested for the cell formation problem considered. Further research is warranted in selecting the appropriate parameter values and hence no guidelines have been provided.

Algorithm

Step 1. Initialization. Select the initial parameters and create an initial diversifed population.

(a) Set the value for PPSZ, XGEN, PCRS, PMUT1, PMUT2 and *C*.
(b) Read the part–machine matrix.
(c) Create an initial population of size PPSZ and call it OLDPOP.
(d) Compute the objective value (weighted sum of voids and exceptional elements, $w = 0.7$) and fitness value (equation 6.1) for each chromosome.
(e) Sort the strings in increasing order of objective value.
(f) Set GEN = 1 (i.e. current generation = 1).

Step 2. Reproduction. Reproduce strings using stochastic sampling without replacement.

(a) Calculate the expected count e_i for each string in OLDPOP.
(b) Allocate samples to a TEMPPOP according to the integer part of e_i and treat the fractional part as success probability.

Step 3. Recombination. Apply recombination operator to TEMPPOP to form a selection pool of population.

(a) Strings to be crossed are selected randomly.
(b) The crossover operator is used sequentially with a probability PCRS. Two chromosomes are chosen randomly to form two new chromosomes.
(c) Apply mutation with a probability of PMUT1.
(d) Calculate the objective value and fitness value for each chromosome.
(e) Sort out the selection pool in increasing order of objective value.

Step 4. Replacement. Compare the chromosomes of sorted OLDPOP and selection pool for their fitness value and create NEWPOP using the replacement policy.

(a) If all the offsprings outperform every existing chromosome in OLDPOP, then all offsprings replace the existing chromosomes in the new population.

(b) If some of them fare better, then replace an equal number of existing chromosomes, i.e. those that are lowest in order of performance in OLDPOP.
(c) For other offsprings, a random selection is made with probability $\beta = 0.005$.

Step 5. Diversification. Apply mutation to diversify the population.

(a) Calculate the diversity parameter H for the current population using equations (6.2) and (6.3).
(b) Compare the diversity with the given acceptable level, execute mutation process repeatedly with probability PMUT2 until the diversity of the population is equal to the acceptable level.
(c) If mutation is performed, calculate the objective value of chromosome.
(d) Sort out the pool of chromosomes in increasing order of objective value.

Step 6. New generation. Evaluate the current generation number to determine the next step.

(a) If GEN $<$ XGEN, then the current population becomes OLDPOP and go to step 2.
(b) If GEN $\geqslant$ XGEN, then stop. The chromosome in the current population with the lowest objective value represents the best solution.

Example 6.2

Consider the data from Example 6.1. The initial parameter values are set as follows: PPSZ = 10, XGEN = 50, PCRS = 0.9, PMUT1 = 0.05, PMUT2 = 1, $C = 2$. The initial generation, first generation and the 50th generation are shown in Table 6.1 for the purpose of illustration. The corresponding objective values of each chromosome in the population are given along with the summary statistics. The best chromosome identified is (2 1 1 2 2 2 1 1). The first four numbers identify the cell to which each machine is assigned. Similarly, the last four numbers identify the part allocation. The part and machine groupings thus obtained are the same as in Example 6.1.

The most difficult issue in the successful implementation of GAs is to find good parameter values. A number of approaches have been suggested to derive robust parameter settings for GAs, including brute-force searches, using mets-level and the adaptive operator fitness technique (Davis, 1991). The optimal parameter values vary from problem to problem. Pakath and Zaveri (1993) proposed a decision-support system to determine the appropriate parameter values in a systematic manner for a given problem.

Table 6.1 Chromosome development

Initial generation Population		*Generation 1 Population*		*Generation 50 Population*	
Chromosome	*Objective*	*Chromosome*	*Objective*	*Chromosome*	*Objective*
12111212	3.3	12111212	3.3	21122211	0.3
22121212	3.3	22121212	3.3	21122211	0.3
22222122	3.5	12111212	3.3	21122211	0.3
22111121	4.3	22121212	3.3	21122211	0.3
22122121	4.3	21121212	3.3	21122212	1.3
21112122	4.7	22222122	3.5	21122212	1.3
12112212	4.7	22222122	3.5	21122212	1.3
21122122	5.3	22222122	3.5	21122212	1.3
12212111	5.3	22122122	3.9	21122212	1.3
21221112	5.7	22122121	4.3	21122212	1.3
Maximum objective value			4.3		1.3
Average objective value			3.52		0.9
Minimum objective value			3.30		0.3
Sum of objective value			35.2		9.0

6.3 NEURAL NETWORKS

Neural network models mimic the way biological brain neurons generate intelligent decisions. Biological brains are superior at problems involving a massive amount of uncertain and noisy data where it is important to extract the relevant items quickly. Such applications range widely, from speech recognition to diagnosis. However, if the problem is well defined and self-contained, traditional serial computing will be superior. Burbridge succinctly stated the main difficulty in solving the cell formation problem by computer as follows (Moon, 1990):

> It is comparatively simple to find groups and families by eye with a small sample. The mental process used combines pattern recognition, the application of production know-how and intuition. However, it has proved to be surprisingly difficult to find a method suitable for the computer which will obtain the same results.

The above experience makes neural network models a potential tool to solve the cell formation problem.

Basically, a neural network consists of a number of processing units linked together via weighted, directed connections (Fig. 6.2). The weights represent the strength of the connections, and are either positive (excitatory) or negative (inhibitory). Each unit receives input signals via weighted incoming connections, then applies a simple linear or non-linear function to the sum of inputs and responds by sending a signal to

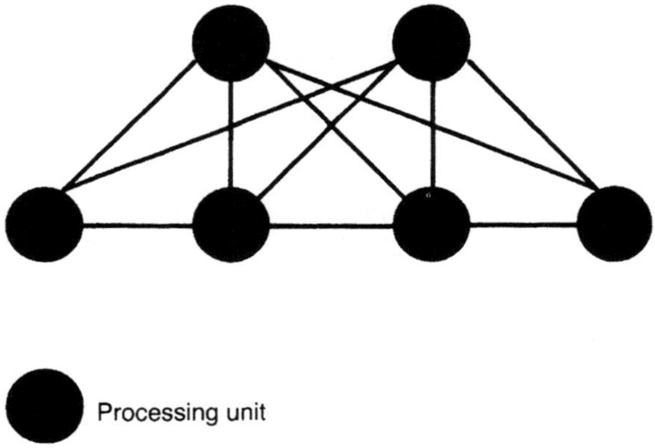

Fig. 6.2 Neural network example.

all of the units to which it has outgoing connections. This basic operation is performed dynamically, concurrently and continuously in every processing unit of the neural network.

There are many neural network models which attempt to simulate various aspects of intelligence. McClelland *et al.* suggested a general framework in which most of these models can be characterized. In this framework, neural network models are suggested to have the following components.

1. Processing units: a biological neuron equivalent. Initial decisions include how many units are needed, how to organize the units and what each unit represents.
2. Pattern of connectivity: specifies how processing units are interconnected and whether the connections are excitatory or inhibitory. Also, each connection is assigned a weight from the pattern information.
3. State of activation: usually takes continuous or discrete values.
4. Activation rule: the output signal values are determined by the activation rule.
5. Output function: determines whether output signals should be generated given the state of activation of each unit.
6. Propagation rule: dictates how to update the activation values of each unit given a new set of connection weights and output signal values from other units.
7. Learning rule: the neural network learns by changing its connection weights and activation values of processing units. The learning rule specifies a systematic modification of such parameters, leading to the modification of connection weights and hence learning.

A particular network model can be considered as a combination of some instances of the above components. The next section adapts the Grossberg's interactive activation and competition network to the cell formation problem.

Interactive activation and competition (IAC) network

An IAC network consists of a collection of processing units organized into a number of competitive pools. There are excitatory connections among units in different pools and inhibitory connections among units within the same pool. The excitatory connections between pools are generally bidirectional, thereby making the processing interactive in the sense that processing in each pool both influences and is influenced by processing in other pools. Within a pool, the inhibitory connections are usually assumed to run from each unit in the pool to every other unit in the pool. This implements a kind of competition among the units such that the unit or units in the pool that receive the strongest activation tend to drive down the activation of other units. The units in an IAC network take on continuous activation values between a maximum and minimum value, although their output, i.e. the signal they transmit to other units, is not necessarily identical to their activation. In this work, the output of each unit tends to be set to the activation of the unit minus the threshold as long as the difference is positive; when the activation falls below threshold the output is set to zero (McClelland and Rumelhart, 1988). The implementation presented in this section was proposed by El-Bouri (1993) and is a modification of the procedure proposed by Moon (1990). The main components of an IAC network are discussed below.

Processing units

Three different pools of processing units are used in this approach. Each pool consists of processing units that represent part types, machine types or cell instances. The number of processing units in the cell instances is either equal to the number of parts or machines. This section considers them equal to the number of parts. The pools for the part types and machine types contain the similarity information among their units through excitatory and inhibitory connections. The cell instances link both the part types and machine types using the information in the part–machine matrix.

Pattern of connectivity

There are two types of connection weight in this network. The first type of weight is between cell instances and part types and between cell instances and machine types, and is given a value of 1 or 0 depending on the information provided in the part–machine matrix. The second type of weight is based on similarity values among machine types and part types which are computed using equations (4.1) and (4.8), respectively.

The weights between unit i and unit j for both part types and machine types are computed using

$$w_{ij} = \begin{cases} s_{ij} - A, & \forall\, i \neq j \\ 0, & \forall\, i = j \end{cases} \tag{6.4}$$

where

$$A = \frac{\sum_i \sum_j s_{ij}}{n(n-1)}, \quad \forall\, i \neq j$$

and s_{ij} is the Jaccards similarity coefficient between units i and j, and n is the number of non-zero entries in the similarity matrix.

State of activation

The state of activation takes continuous values less than unity 1. The magnitude of the value indicates the strength with which it interacts with a specific unit.

Activation rule and output function

Each processing unit receives an external input from the connected units and modifies its current activation accordingly. The new activation influences the input to adjacent units and the effect propagates through the network until a stable state is reached. A combined input to a processing unit i is calculated as follows:

$$\text{netinput}\ (i) = \sum_j w_{ij}\ \text{output}\ (j)\ + \text{extinput}\ (i)$$

where

$$\text{output}\ (j) = [\text{act}\ (j)]^+$$

and

$$[\text{act}\,(j)]^+ = \begin{cases} \text{act}(j), & \text{if act}\ (j) > 0 \\ 0, & \text{otherwise} \end{cases}$$

The activation values are updated according to the following:

$$\text{act}(i) = \begin{cases} \text{act}(i) + \text{netinput}(i)(\max - \text{act}(i)) - \text{decay}(\text{act}(i) - \text{rest}), & \text{if netinput}(i) > 0 \\ \text{act}(i) + \text{netinput}(i)(\text{act}(i) - \min) - \text{decay}(\text{act}(i) - \text{rest}) & \text{otherwise} \end{cases}$$

where max = 1; min $\leqslant$ rest $\leqslant$ 0; and 0 < decay < 1. The decay rate determines how quickly the stable condition is reached. The stable condition is when the activation values of all the processing units do not vary by more than 0.1% of the previous cycle. If the decay rate is decreased, the execution of the network slows down. On the other hand, too high a decay rate may result in an oscillatory situation in which the resting level is never achieved. The values are chosen by trial and error to be between 0.05 and 0.2.

Propagation rule

The propagation rule controls the network change of state. Processing units are picked randomly and updated once. When all units have been updated, one cycle is said to be completed and a new one started. This processing continues until stability is reached after a number of cycles. Stability is reached when the state of activation does not change any further.

Learning rule

In an IAC network, the connection weights are set a priori and do not change. Thus, this network does not use any learning rules.

Parameter values

In the IAC model, as run on PDP software, there are several parameters under the user's control:

- max, the maximum activation parameter;
- min, the minimum activation parameter;
- rest, the resting activation level to which activations tend to settle in the absence of external input;
- decay, determines the strength of the tendency to return to resting level;
- estr, scales the influence of external signals relative to internally generated inputs to units;
- alpha, scales the strength of the excitatory inputs to units from other units in the network;
- gamma, scales the strength of inhibitory inputs to units from other units in the network.

These parameters control the pace and magnitude of the interaction between the processing units. The most important parameters are alpha and gamma, which scale the excitatory and inhibitory connections, respectively, between the units. A higher gamma stresses the inhibitory connections leading to many small cells, while a higher alpha increases the effects of the excitatory connections and identifies large, loose cells.

Neural network algorithm using IAC

The algorithm proposed by El-Bouri (1993) is presented in this section. The algorithm requires the use of PDP software developed by McClelland and Rumelhart (1988). Whenever a unit is clamped and the network is run to stability, several other units are usually activated positively or negatively, with varying degrees of activation. When a unit is said to be clamped, it is forced to be selected in the group. A threshold value needs to be defined to restrict the number of units joining a group. A low threshold value will lead to the identification of few large cells. On the other hand, a high threshold value will identify many small, tight cells. Depending on the unit clamped, the other units may be assigned to one or more clamped units, thus requiring the decision-maker to allocate these units to cells. This situation is encountered when a unit, say A, has a strong connection with two units B and C; however, B and C are not strongly connected. Unit A will then be referred to as a 'double assignment'. In the algorithm proposed, all units that cannot be uniquely assigned to a cell are left until the end and assigned to the cell using a scoring scheme. This avoids deciding on the threshold value, which is often subjective. The algorithm is designed to compromise between the two conflicting objectives: minimizing the number of exceptional elements and voids. Step 2 in the algorithm below promotes grouping machines such that it results in a minimum number of exceptional elements, while step 9 attempts to manipulate assignments to result in fewer voids.

Step 1. Select any unassigned machine at random and clamp it. If all the machines are assigned, go to step 5.
Step 2. Run the IAC using PDP software until the network has reached stability. Let machine A be clamped and say machine S has an activation close to that of A.

(a) Assign machine S to machine A if it has not already been assigned to any group. In case of a tie, assign all qualifying machines.
(b) If machine S is already assigned to a group and machine A is the sole member of the current group, assign A to the cell containing S. Return to step 1, else go to (c).
(c) Remove machine S and keep it in a separate list of doubly-assigned machines.

Step 3. Clamp machine S or any newly assigned machine (if a tie existed) in step 2, and go to step 2.

Step 4. If all the machines in step 2 have been clamped once, close the machine group and return to step 1.
Step 5. If the list of doubly-assigned machines is empty go to step 7, else go to step 6.
Step 6. Clamp the first machine on the list of doubly-assigned machines and run the IAC.

(a) Find the sum of positive activations for each of the existing groups.
(b) Assign the clamped machine to the group with the highest score; go to step 5.

Step 7. Clamp all the machines assigned to a group simultaneously and run the IAC.

(a) Assign all the positively activated part units to the same cell as the machines. All double-assigned parts are placed in a separate list.
(b) Repeat steps 7 and 7(a) for all machine groups.
(c) If a few parts are unassigned, place them in the list of double-assigned parts.

Step 8. If the list of double-assigned parts is empty, go to step 10, else go to step 9.
Step 9. Clamp the first part on the list of double-assigned parts and run the IAC.

(a) Count the total number of positively activated units for each of the existing groups.
(b) Assign the clamped machine to the group with the highest score. In case of a tie, assign it to the group with fewer machines; go to step 8.

Step 10. Stop.

The above algorithm can also be run by clamping parts first instead of machines. It is suggested the decision to clamp machines or parts should be made depending on whichever is less numerically.

Example 6.3

Consider the data from Example 6.1. The parameters are set as follows: max = 1, min = − 0.2, rest = − 0.1, decay = 0.1, estr = 0.01, alpha = 0.15, gamma = 0.15 and number of cycles = 50 (to reach stability). In constructing the network, 12 processing units are needed: four for part instances, four for machine instances and four for cell instances. The connection weights between units in different pools are given a value of 0 or 1 depending on the relationship indicated by the part–machine matrix. For units within each pool (part and machine) the weights are computed using the similarity matrix and equation (6.4). The similarity matrices and weight matrix are shown in Figs. 6.3–6.5. The complete network with the connection weights is shown in Fig. 6.6.

Step 1. Clamp machine 1.
Step 2. After running the IAC to stability, the following activations are obtained:

*m1	0.91	p1	0.44
m2	−0.16	p2	0.44
m3	−0.16	p3	−0.14
m4	0.62	p4	−0.14

Step 2(a) applies, therefore cell 1 contains m1 and m4.
Step 3. Clamp m4, go to step 2.
Step 2. After running the IAC to stability, the following activations are obtained:

m1	0.62	p1	0.44
m2	−0.16	p2	0.44
m3	−0.16	p3	−0.14
*m4	0.91	p4	−0.14

Step 4. applies since all the machines in cell 1 have been clamped; return to step 1.
Step 1. Clamp m2.
Step 2. After running the IAC to stability, the following activations are obtained:

	m1	m2	m3	m4
m1	0	0	0	0.5
m2	0	0	0.5	0
m3	0	0.5	0	0
m4	1	0	0	0

Fig. 6.3 Similarity matrix for machines.

	p1	p2	p3	p4
p1	0	1	0	0
p2	1	0	0	0
p3	0	0	0	0.5
p4	0	0	0.5	0

Fig. 6.4 Similarity matrix for parts.

	m1	m2	m3	m4	p1	p2	p3	p4	c1	c2	c3	c4
m1	0	0	0	0.25	0	0	0	0	1	1	0	0
m2	0	0	−0.25	0	0	0	0	0	0	0	1	1
m3	0	−0.25	0	0	0	0	0	0	0	0	1	0
m4	0.25	0	0	0	0	0	0	0	1	1	0	0
p1	0	0	0	0	0	0.25	0	0	1	0	0	0
p2	0	0	0	0	0.25	0	0	0	0	1	0	0
p3	0	0	0	0	0	0	0	−0.25	0	0	1	0
p4	0	0	0	0	0	0	−0.25	0	0	0	0	1
c1	1	0	0	1	1	0	0	0	0	−2	−2	−2
c2	1	0	0	1	0	1	0	0	−2	0	−2	−2
c3	0	1	1	0	0	0	1	0	−2	−2	0	−2
c4	0	1	0	0	0	0	0	1	−2	−2	−2	0

Fig. 6.5 Weight matrix for the network.

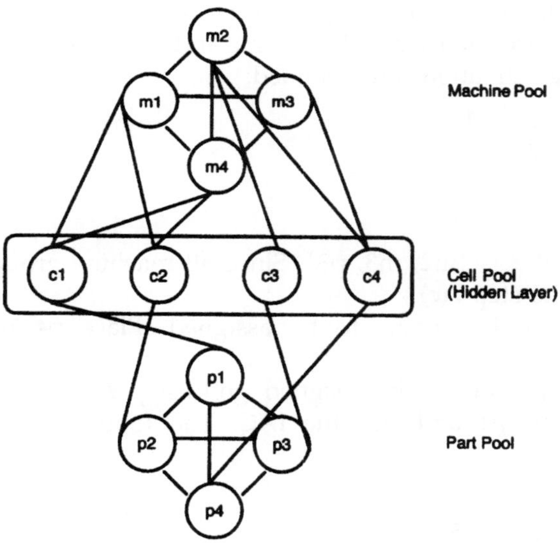

Fig. 6.6 Complete network.

m1	−0.15	p1	−0.13
*m2	0.90	p2	−0.13
m3	0.35	p3	0.46
m4	−0.15	p4	−0.11

Step 2(a) applies, therefore cell 2 contains m2 and m3.
Step 3. Clamp m3, go to step 2.
Step 2. After running the IAC to stability, the following activations are obtained:

m1	−0.15	p1	−0.13
m2	0.35	p2	−0.13
*m3	0.90	p3	0.46
m4	−0.15	p4	−0.11

Step 4 applies since all the machines in cell 2 have been clamped; return to step 1.
Step 1. Since all the machines have been clamped once, go to step 5.
Step 5. Since there are no doubly-assigned machines, go to step 7.
Step 7. Clamp machines in cell 1, i.e. m1 and m4. After running the IAC, the following activations are obtained:

*m1	0.91	p1	0.46
m2	−0.16	p2	0.46
m3	−0.16	p3	−0.15
*m4	0.91	p4	−0.15

Step 7(a). Cell 1 is {m1, m2, p1, p2}; go to step 7.
Step 7. Clamp machines in cell 2, i.e. m2 and m3. After running the IAC, the following activations are obtained:

m1	−0.16	p1	−0.13
*m2	0.90	p2	−0.13
*m3	0.90	p3	0.48
m4	−0.16	p4	−0.11

Step 7(a). Cell 2 is {m2, m3, p3}. Since all machine groups have been clamped, go to step 7(c).
Step 7(c). Part 4 remains to be assigned. Place p4 in the list of unassigned parts.
Step 8. Since p4 is yet to be assigned, go to step 9.
Step 9. Clamp p4 and run the IAC. The following activations are obtained:

m1	−0.13	p1	−0.15
m2	0.43	p2	−0.15
m3	−0.11	p3	−0.12
m4	−0.13	p4	0.90

Step 9(a). Since the number of positively activated units is none for cell 1 and one for cell 2, p4 is assigned to cell 2. Cell 2 is {m2, m3, p3, p4}. Go to step 8.
Step 8. Since no unassigned parts in list, go to step 10.
Step 10. Stop.

Note that the above is a very simple problem and the solution could have been obtained at step 1 by grouping all positively activated machines and parts in one cell, removing them from the list and then repeating this step for m2 or m3. However, this procedure as proposed by Moon (1990) does not provide a systematic approach to double assignments and need not necessarily minimize the voids and exceptional elements in the solution.

6.4 RELATED DEVELOPMENTS

A review and comparison of these emerging heuristics was provided by Glover and Greenberg (1989). Holland (1992) and Masson and Wang (1990) provide an excellent introduction to genetic algorithms and neural networks, respectively. In the context of cell formation, Boctor (1991) applied simulated annealing to the machine grouping problem. The problem he considered minimizes the number of exceptional elements with constraints imposed on cell sizes. Venugopal and Narendran (1992) implemented a genetic algorithm considering the two objectives of minimizing the volume of inter-cell moves and minimizing the total within cell load variations. Moon (1990) adopted the constraint satisfaction model of neural networks with the flexibility of the similarity coefficient method. The sequence of operations, alternate process plans and lot size were considered in the implementation. Rao and Gu (1992) presented a self-organizing neural network, called the adaptive resonance theory (ART1). A modified ART1 is then applied for part–machine grouping. The number of articles published in these areas is growing fast (Moon and Chi, 1992; Schaffer *et al.*, 1989; Pakath and Zaveri, 1993).

6.5 SUMMARY

The heuristics presented in previous chapters are iterative and are sensitive to the initial solution and data set. The general disadvantages of such iterative algorithms are as follows (Laarhoven and Aarts, 1987):

- by definition, iterative algorithms terminate in a local minimum and there is generally no information on the amount by which this local minimum deviates from a global minimum;
- the obtained local minimum depends on the initial configuration, for the choice of which no guidelines are generally available;
- in general it is not possible to give an upper bound for the computation time.

However, it should be clear that the iterative algorithm can on average be executed in a small computation time. One way to avoid some of the above disadvantages is to execute the iterative algorithm for a large number of initial solutions.

Random search algorithms such as simulated annealing, genetic algorithms and neural networks provide solutions which do not depend on the initial solution and have an objective value closer to the global optimum value. However, the gain in applying these general algorithms can sometimes be undone by the computational effort, since these procedures are slower than the iterative algorithms. In this chapter, each of these procedures was introduced in the context of cell formation. The powers and limitations of SA, GAs and NNs remains to be researched. In spite of many published expositions (referring to other areas of research), including mathematical convergence proofs under special assumptions, the fundamental scope and supporting rationale for these approaches are not completely known. Current research is unable to explain adequately why or when these approaches are likely to succeed or fail (Glover and Greenberg, 1989).

PROBLEMS

6.1 Briefly discuss the inspiration behind the development of simulated annealing, genetic algorithms and neural networks.

6.2 Why are these algorithms referred to as the random search algorithms?

6.3 Why have these procedures gained the attention of researchers? What are the possible application areas?

6.4 Write computer code to implement the simulated annealing algorithm and obtain the groupings for the part–machine matrix in Q 5.1. Experiment with the values of T_0 and α.

6.5 Implement the simulated annealing algorithm to consider alternate process plans. Test your procedure on the data given in Q 5.7. Compare the solution obtained with the optimal solution and the iterative procedure.

6.6 Write computer code to implement the genetic algorithm and obtain the groupings for the part–machine matrix in Q 5.1. Experiment with the values of PPSZ and XGEN.

6.7 Using the IAC algorithm obtain the groupings for the part–machine matrix in Q 5.1. Experiment with the values of alpha and gamma.

REFERENCES

Adil, G.K., Rajamani, D and Strong, D. (1994) Simulated annealing algorithm for part machine grouping. Univ. of Manitoba, Canada. Working paper.

Baker, J.E. (1987) Adaptive selection methods for genetic algorithms, in *Proceedings of the 1st International Conference on Genetic Algorithms and Applications.* (ed. J.J. Grefenstette), Lawrence Erlbaum Associates, pp. 14–21.

Boctor, F.F. (1991) A linear formulation of the machine part cell formation problem. *International Journal of Production Research*, **28**(1), 185–98.

Booker, L.B. (1987) Improving search in genetic algorithms, in *Genetic Algorithms and Simulated Annealing*, (ed. L. Davis), Morgan Kaufmann Publishers, Los Angeles, pp. 61–73.

Davis, L. (1991) *Handbook of Genetic Algorithms*, Edited. Van Nostrand Reinhold, New York.

De Jong, K.A. (1975) An analysis of the behavior of a class of genetic adaptive systems. Univ. Michigan. Doctoral dissertation. (*Dissertation Abstracts International*, **36**(10) 5140B).

El-Bouri, A. (1993) Part machine groupings using interative activation and competition networks. Univ. Manitoba, Canada. Working paper.

Francis, R.L., McGinnis, L.F. and White, J.A. (1992) *Facility Layout and Location: an Analytical Approach*, Prentice-Hall, Englewood Cliffs, NJ.

Glover, F. and Greenberg, H. J. (1989) New approaches for heuristic search: a bilateral linkage with artificial intelligence. *European Journal of Operational Research*, **39**, 119–30.

Goldberg, D.E. (1989) *Genetic Algorithms in Search, Optimization and Machine Learning*, Addison-Wesley, Reading, MA.

Grefenstette, J.J. (1987) Incorporating problem specific knowledge into genetic algorithms and simulated annealing, in *Genetic Algorithms and Simulated Annealing*, (ed. L. Davis), Morgan Kaufmann Publishers, Los Angeles.

Gupta, Y., Gupta, M., Kumar, A. and Sundaram, C. (1994) A genetic algorithm based approach to cell composition and layout design problems. Univ. Louisville, KY. Working paper.

Gupta, M. and Rajamani, D. (1994) A genetic algorithm to the cell formation problem. Univ. Manitoba, Canada. Working paper.

Holland, J. (1992) Genetic algorithms. *Scientific American*, July, 66–72.

Van Laarhoven, P.J.M. and Aarts, E.H.L. (1987) *Simulated Annealing: Theory and Applications*, Kluwer Academic Publications, Dordrecht, Netherlands.

Liepins, G.E. and Hillard, M.R. (1989) Genetic algorithms: foundations and applications. *Annals of Operations Research*, **21**, 31–58.

Masson, E. and Wang, Y. J. (1990) Introduction to computation and learning in artificial neural networks. *European Journal of Operational Research*, **47**, 1–28.

McClelland, S.L. and Rumelhart, D.E. (1988) *Explorations in Parallel Distributed Processing: a Handbook of Models, Programs and Exercises*, MIT Press, Cambridge, MA.

Moon, Y.B. (1990) Forming part families for cellular manufacturing: a neural network approach. *International Journal of Advance Manufacturing Technology*, **5**, 278 –91.

Moon, Y.B. and Chi, S.C. (1992) Generalized part family formation using neural network techniques. *Journal of Manufacturing Systems*, **11**(3), 149–59.

Pakath, R. and Zaveri, J.S. (1993) Specifying critical inputs in a genetic s-driven decision support system: an automated facility. Univ. Kentucky. Working paper.

Rao, H.A. and Gu, P. (1992) Design of cellular manufacturing systems. *International Journal of Systems Automation: Research and Applications*, **2**, 407–24.

Schaffer, J.D., Caruana, R.A., Eshelman, L.J. and Das, R. (1989) A study of control parameters affecting online performance of genetic algorithms for function optimization, in *Proceedings of the 3rd International Conference on Genetic Algorithms*, pp. 51–60.

Venugopal, V. and Narendran, T.T. (1992) A genetic algorithm approach to the machine component grouping problem with multiple objectives. *Computers and Industrial Engineering*, **22**(4), 469–80.

CHAPTER SEVEN

Other mathematical programming methods for cell formation

A number of cells are created using new and automated machines and material handling equipment. Flexible manufacturing systems (FMS) are examples of automated cells where production control activities are under computer supervision. Considering the versatility of these machines and high capital investment, a judicious selection of processes and machines is necessary during cell formation. The creation of independent cells, i.e. cells where parts are completely processed in the cell and there are no linkages with other cells, is a common goal for cell formation. If it is assumed that there is only one unique plan (operation sequence and machine assignment) for each part, then the creation of independent cells may not be possible without duplication of machines. The duplication of machines requires additional capital investment. However, if we allow for alternate plans (operation sequence and machine assignment), and for each part select plans during cell formation, it may be possible to select plans which can be processed within a cell without additional investment.

However, on many occasions it may not be economical or practical to achieve cell independence. Allowing for alternate plans may also lead to cost reduction in inter-cell material handling movement. Therefore, it is important to integrate the essential factors and consider the economics of these aspects during cell design. Moreover, with the introduction of new parts and changed demands, new part families and machine groups have to be identified if cells have already been established. The redesign of such systems warrants the consideration of practical issues such as the relocation expense for existing machines, investment on new machines etc. In fact, new technology and faster deterioration rates of certain machines could render the previously allocated parts and machines undesirable. Thus, there is also need to determine whether the old machines must be replaced with new or technologically updated machines. This chapter provides a mathematical framework for addressing many of these issues and discusses the relevant literature in this area.

Cell formation as defined in this chapter, in addition to identifying the part families and machine groups, specifies the plans selected for each part, the quantity to be produced through the plans selected, the machine type to perform each operation in the plans, the total number of machines required, and machines to be relocated (for redesign) considering demand, time, material handling and resource constraints. Some pertinent objectives to be considered are minimization of investment, operating cost, machine relocation cost and material handling cost. Considerations of physical limitations such as the upper bound on cell size, machine capacity and material handling capacity are also incorporated in the cell formation process.

7.1 ALTERNATE PROCESS PLANS

Process planning is the systematic determination of the methods by which a product is manufactured economically and competitively. A process planner examines each part to identify the sequence of operations required to transform the raw material to a finished part. The machines can be classified into different types according to the operations they can perform. An operation can be performed on different types. Machines are selected for each operation taking into consideration other operational aspects such as the existence of fixtures, machine capacities, quality of parts etc. Once the operations and machines are matched, the part–machine matrix is generated. In previous chapters the part families and machine groups were identified based on the routings thus identified. Note that these routings were prepared independently of each other, with no reference to cell formation. As pointed out in Chapter 5, considering alternate plans for parts and interaction with other plans can lead to a better cell production system. It can be realized that a simple part made in five operations on machine tools of which there are, say, four different machine types that can be used (any might be employed for an operation) will have over 1000 possible routes. Possible changes in the sequence of operations may increase this figure (Burbidge, 1992). If this enormous variability in route data is considered, it is possible to obtain a total division of the factory into independent cells.

For example, consider the manufacture of a gear (Rajamani, Singh and Aneja, 1990). If the initial raw material is in the form of a bar stock, the following eight processing steps (PS) are required to transform the raw material into a finished gear:

PS 1: facing
PS 2: turning
PS 3: parting-off

PS 4: facing
PS 5: centering
PS 6: drilling
PS 7: slotting
PS 8: gear teeth cutting

A different set of processing steps can be identified if the raw material is in a different form, blanks say, either cast or forged. Once the processing steps have been identified, the process planner determines the possible processing sequences before grouping the processing steps into operations. The eight processing steps in the gear manufacture can be grouped into different sets as shown in Table 7.1.

It is possible to alter such a grouping to suit the manufacturing system requirements. For example, the first six processing steps can be combined to perform them with one setup, say, on a turret lathe. Further, PS 6 can be separated and performed on a drilling machine. Also, each operation in the plans can be performed on a number of compatible machines. For example, the gear teeth cutting operation can be performed either on a milling or a gear hobbing machine if plan 1 is used. If plan 2 is used, where the gear teeth cutting and slotting operations have been combined, it can only be performed on a milling machine. The next section analyses how alternate process plans influence the resource utilization when part families and machine groups are formed concurrently.

7.2 NEW CELL DESIGN WITH NO INTER-CELL MATERIAL HANDLING

The ultimate application of GT in manufacturing is the formation of mutually exclusive cells. A number of heuristics used to form part families and machine groups have been discussed in earlier chapters. In these procedures it was assumed that each part can be produced using one process plan. Also, if a machine was assigned to a cell, it was assumed sufficient capacity was available in each cell to process all the parts assigned to the cell. Moreover, all copies of the same machine type were lumped and considered as one machine. Manufacturing aspects

Table 7.1 Grouping processing steps into operations

Operation	*Plan 1*	*Plan 2*
1	PS 1, 2, 3	PS 1, 2, 3
2	PS 4, 5, 6	PS 4, 5, 6
3	PS 7	PS 7, 8
4	PS 8	

such as demand, processing cost etc., were not considered in this process. In this section we are interested in developing new cells for the parts to be produced. Thus, the objective is to identify part families and machine groups by selecting the appropriate process plans for each part and machine type such that the total investment cost is minimized. This can be accomplished by adopting either a sequential approach or a simultaneous approach. In the sequential approach, the part families will be identified based on part attributes and machines will be assigned to part families such that the parts can be exclusively processed within a cell. In the simultaneous approach, part families and machine groups will be identified concurrently. Accordingly, in the following subsections two mathematical models for optimal cell design will be presented (Rajamani, 1990). The assumptions and notation are stated before presenting the models.

Assumptions

This chapter assumes that a part can be produced through one or more process plans. A process plan for a part consists of a set of operations. Each operation in a process plan can be performed on alternate machines (Rajamani, 1990). Thus for each process plan we have a number of plans depending on the machines selected for each operation. These plans are referred to as production plans. For example, say a process plan for a part requires two operations and each operation can be performed on two types of machine. There are four different production plans which can be used to produce the part. It is also assumed that the demand for a part could be split and can be produced in more than one cell. In the above example, one or more of the four production plans can be used together to produce the part. The plans identified to produce the part in different cells could be different.

Notation

b_m time available on each machine of type m
C_m cost of each machine of type m
$c = 1, 2, \ldots\ldots C$ cells
d_k demand for part k
k $1, 2, \ldots\ldots K$ parts
$L = 1, 2, \ldots L$ (kp) production plans for (kp) combinations
$m = 1, 2, \ldots M$ machines
$p = 1, 2, \ldots P_k$ process plans for part k
$s = 1, 2 \ldots S\,(kp)$ operations for (kp) combinations
$X(lkp)$ amount of part k produced using process plan p and production plan l

$$a_{ms}(lkp) = \begin{cases} 1, & \text{if in plan } l \text{ machine } m \text{ is assigned to operation } s \text{ for all } (kp) \\ 0, & \text{otherwise} \end{cases}$$

$$c_{ms}(kp) = \begin{cases} \text{cost for machine } m \text{ to perform operation } s \text{ for all } (kp) \\ \infty, \quad \text{if machine } m \text{ cannot perform this operation} \end{cases}$$

$$t_{ms}(kp) = \begin{cases} \text{time for machine } m \text{ to perform operation } s \text{ for all } (kp) \\ \infty, \quad \text{if machine } m \text{ cannot perform this operation} \end{cases}$$

Other variables and parameters will be defined when required.

Sequential grouping model

This model adopts a sequential approach to cell formation by assigning machines to known part families. This model has wide applicability because a number of companies have indicated the use of one or more classification schemes in conjunction with GT applications for determining part families. The part families thus formed were determined without reliance on production methods (Wemmerlov and Hyer, 1989). Information on which part belongs to which part family is denoted by an indicator β_{kf},

$$\beta_{kf} = \begin{cases} 1, & \text{if part } k \text{ belongs to part family } f \\ 0, & \text{otherwise} \end{cases}$$

Also let Z_{mf} be the number of machines of type m used to produce parts in family f. Accordingly, the model is as described below.

Machine grouping model

Minimize

$$\sum_{mf} C_m Z_{mf} + \sum_{kpl} \left(\sum_{ms} a_{ms}(lkp)\, c_{ms}(kp) \right) X(lkp)$$

subject to

$$\sum_{pl} X(lkp) \geqslant d_k \quad \forall k \tag{7.1}$$

$$\sum_{k} \beta_{kf} \sum_{pl} \left(\sum_{s} a_{ms}(lkp)\, t_{ms}(kp) \right) X(lkp) \leqslant b_m Z_{mf}, \quad \forall m,f \tag{7.2}$$

$$Z_{mf} \text{ have non-negative integer values, } \forall\, m,f; X(lkp) \geqslant 0, \forall\, l,k,p \tag{7.3}$$

Constraints 7.1 guarantee that the demand for all part types is met. Constraints 7.2 ensure that the capacity of each machine type assigned to all the part families is not violated. The integer variables are indicated by constraints 7.3. Since the number of machines of each type selected is restricted to be a non-negative integer, the model implicitly minimizes the under-utilization of machines.

In the above formulation, each production plan l for part k and process plan p is an assignment of machines to each of the $S(kp)$ operations. The set of all such possible production plans is denoted by $L(kp)$. Thus a plan $l(\varepsilon)L(kp)$ is defined by an assignment given by:

$$a_{ms}(lkp) = \begin{cases} 1, & \text{if in plan } l \text{ machine } m \text{ is assigned to operation } s \text{ for all } (kp) \\ 0, & \text{otherwise} \end{cases}$$

These numbers are only known implicitly in the model developed. Since $L(kp)$ is large, we have a large number of columns. However, the model can be solved via a column generation procedure. The implicitly known columns $(X(lkp))$ are generated through a greedy procedure since the corresponding optimization procedure turns out to be a specialized assignment problem. The column generation procedure will be presented in the next section. This section enumerates all possible plans and solves the model to illustrate the impact of alternate process plans.

Example 7.1

Four different part types of known demand ($d_1 = d_2 = d_3 = d_4 = 100$) are manufactured, each with 2, 2, 3 and 2 process plans, as given in Table 7.2. Each operation in a plan can be performed on alternate machines. Three types of machines of known capacity ($b_1 = b_2 = b_3 = 1000$) and discounted cost ($C_1 = 1000$; $C_2 = 2500$; $C_3 = 3000$) are available. The time and cost information for performing an operation on compatible machines for each process plan is given in Table 7.3, which also explicitly enumerates all possible production plans. To solve the sequential model it is assumed parts 1 and 2 belong to the first part family and parts 3 and 4 are identified as the second part family. The results obtained by solving the model are given in Table 7.4.

Table 7.2 Time and cost required for machine m to perform operation s on part k using process plan p

		$k=1$		$k=2$		$k=3$			$k=4$	
		$p=1$	$p=2$	$p=1$	$p=2$	$p=1$	$p=2$	$p=3$	$p=1$	$p=2$
$s=1$	$m=1$	5,3		3,4		2,2		8,1	1,2	9,7
	$m=3$	7,2		4,3		2,2		9,2	2,1	8,9
$s=2$	$m=2$	3,5	9,8	7,8	3,3	3,3	1,2	5,9	2,3	9,8
	$m=3$	4,3	7,9	7,7	2,3	4,4	2,4	3,10	2,4	10,9
$s=3$	$m=1$		8,8	10,9	6,5		11,7	7,4	3,5	
	$m=2$		7,7	8,9	6,6		8,8	9,5	2,6	

Table 7.3 Time and cost information for different production plans

			$m=1$	$m=2$	$m=3$
$k=1$	$p=1$	$l=1$	5,3	3,5	
		$l=2$	5,3		4,3
		$l=3$		3,5	7,2
		$l=4$			11,5
$k=1$	$p=2$	$l=1$	8,8	9,8	
		$l=2$		16,15	
		$l=3$	8,8		7,9
		$l=4$		7,7	7,9
$k=2$	$p=1$	$l=1$	13,13	7,8	
		$l=2$	3,4	15,17	
		$l=3$	13,13		7,7
		$l=4$	3,4	8,9	7,7
		$l=5$	10,9	7,8	4,3
		$l=6$		15,17	4,3
		$l=7$	10,9		11,10
		$l=8$		8,9	11,10
$k=2$	$p=2$	$l=1$	6,5	3,3	
		$l=2$		9,9	
		$l=3$	6,5		2,3
		$l=4$		6,6	2,3
$k=3$	$p=1$	$l=1$	2,2	3,3	
		$l=2$	2,2		4,4
		$l=3$		3,3	2,2
		$l=4$			6,6
$k=3$	$p=2$	$l=1$	11,7	1,2	
		$l=2$		9,10	
		$l=3$	11,7		2,4
		$l=4$		8,8	2,4
$k=3$	$p=3$	$l=1$	15,5	5,9	
		$l=2$	8,1	14,14	
		$l=3$	15,5		3,10
		$l=4$	8,1	9,5	3,10
		$l=5$	7,4	5,9	9,2
		$l=6$		14,14	9,2
		$l=7$	7,4		12,12
		$l=8$		9,5	12,12
$k=4$	$p=1$	$l=1$	4,7	2,3	
		$l=2$	1,2	4,9	
		$l=3$	4,7		2,4
		$l=4$	1,2	2,6	2,4
		$l=5$	3,5	2,3	2,1
		$l=6$		4,9	2,1
		$l=7$	3,5		4,5
		$l=8$		2,6	4,5
$k=4$	$p=2$	$l=1$	9,7	9,8	
		$l=2$	9,7		10,9
		$l=3$		9,8	8,9
		$l=4$			18,18

Simultaneous grouping model

To develop the sequential model it was assumed that the part families were known. This method may not uncover natural part familes, because the part familes were based on part attributes and not manufacturing attributes. Moreover, not all companies have well-developed coding schemes to establish part families. To discover natural part families and machine groups, they have to be formed simultaneously. To develop this model the decision variable is defined to reflect the fact that the demand for a part can be met in more than one cell. The plans selected to produce the parts in the respective cells may be different but they are exclusively processed in that cell with no inter-cell movement. Since in many practical situations, physical floor space restricts the maximum number of machines in each cell, this additional information is required. The model is given by:

Minimize

$$\sum_{mc} C_m Z_{mc} + \sum_{kpcl} \left(\sum_{ms} a_{ms}(lkpc)\, c_{ms}(kp) \right) X(lkpc)$$

subject to:

$$\sum_{cpl} X(lkpc) \geq d_k, \quad \forall k \tag{7.4}$$

Table 7.4 Solution for the sequential model (objective Function value = 10116.67).
(a) Indicates the process plan P and production plan l selected for each part.

			$s=1$	$s=2$	$s=3$	$X(lkp)$
$k=1$	$p=1$	$l=1$	$m=1$	$m=2$		100
$k=2$	$p=2$	$l=1$		$m=1$	$m=2$	83.3*
	$p=2$	$l=2$		$m=2$	$m=3$	16.67
$k=3$	$p=1$	$l=1$	$m=1$	$m=2$		100
$k=4$	$p=1$	$l=1$	$m=1$	$m=2$	$m=1$	100

(b) Optimum number of each machine type assigned to each part family

	Part family $f=1$	Part family $f=2$
$m=1$	1	1
$m=2$	1	1
$m=3$	0	0

*rounded to an integer with a small increase in cost.
(Source: Rajamani, Singh and Aneja (1990); reproduced with permission from Taylor and Francis)

$$\sum_{kpl}\left(\sum_{s} a_{ms}(lkpc)\, t_{ms}(kp)\right) X(lkpc) \leqslant b_m Z_{mc}, \quad \forall m,c \tag{7.5}$$

$$\sum_{m} Z_{mc} \leqslant Max_c, \quad \forall c \tag{7.6}$$

$$Z_{mc} \text{ have non-negative integer values, } \forall m,c; X(lkpc) \geqslant 0, \forall l,k,p,c \tag{7.7}$$

where $L(kpc)$ is the number of different production plans for (kpc) combinations,

$$a_{ms}(lkpc) = \begin{cases} 1, & \text{if in plan } l \text{ machine } m \text{ is assigned to operations for all } (kpc) \\ 0, & \text{otherwise} \end{cases}$$

Max_c = maximum number of machines in cell c

$X(lkpc)$ is the amount of part k produced using process plan p and production plan l in cell c, and Z_{mc} is the number of machines of type m in cell c; and constraints 7.4, 7.5 and 7.7 correspond to constraints 7.1–7.3, respectively. The maximum number of machines which can be assigned to each cell are imposed by constraints 7.6. The value of C ($c = 1,2\ldots C$) denotes that the number of cells is based on judgement. It is suggested that this value can be an overestimate. Only the required number of cells will be formed leaving the other cells empty.

Example 7.2

Consider the same input as provided for Example 7.1. In this case, however, the information on part families is not required and will be determined by the model optimally. Additional information on the number of cells and the maximum number of machines in each cell is needed for this model. It is assumed that two cells have to be formed with a physical limitation of not more than two machines in each cell. The solution to the simultaneous grouping model is provided in Table 7.5.

The results shown in Tables 7.4 and 7.5 indicate that both models have selected the same process plans for each part. However, the objective function values are 10116.67 and 9233.33, respectively. The presence of alternate machines and the simultaneous grouping of part families and machine groups is the main reason for these resource savings. This is indicated by the different production plan selection for parts. Moreover, the simultaneous grouping model identifies natural part families based on manufacturing attributes which otherwise are assumed to be known in the sequential grouping model. This indicates that the sequential approach to forming part families and assigning machines to part families can lead to inferior performance in terms of

Table 7.5 Solution for the simultaneous model (objective function value = 9233.33).
(a) Indicates the process plan P and production plan l selected for each part.

			$s=1$	$s=2$	$s=3$	$X(lkpc)$
$k=1$	$p=1$	$l=1$	$m=1$	$m=2$		100
$k=2$	$p=2$	$l=2$		$m=2$	$m=2$	100
$k=3$	$p=1$	$l=1$	$m=1$	$m=2$		100
$k=4$	$p=1$	$l=1$	$m=1$	$m=2$	$m=1$	66.67*
	$p=1$	$l=2$	$m=1$	$m=2$	$m=2$	33.33

(b) Optimum number of each machine type assigned to each part family. Parts 1,3 and 4 are assigned to cell 1; part 2 is assigned to cell 2.

	Cell $c=1$	*Cell* $c=2$
$m=1$	1	0
$m=2$	1	1
$m=3$	0	1

*rounded to an integer with a small increase in cost.
(Source: Rajamani, Singh and Aneja (1990); reproduced with permission from Taylor and Francis)

resource utilization. Although this observation is made with one cost vector, the same conclusion could be drawn for any other cost vector. This follows from noting that any solution, specifically the optimal solution using the sequential approach, provides a feasible solution for the simultaneous approach. Thus, the simultaneous grouping model would provide results at least as good as the sequential model.

7.3 NEW CELL DESIGN WITH INTER-CELL MATERIAL HANDLING

The creation of independent cells is a common goal for cell formation. However, in many situations it may not be economical or practical to achieve cell independence, especially when under-utilization, load imbalance and higher capital investment are the potential problems of introducing cellular manifacturing. Therefore, there is a need to consider the material handling cost arising during inter-cell movement in these circumstances (Rajamani, Singh and Aneja, 1993). To introduce the material handling cost between cells the production plan definition should contain information on not only the machine type but also the cell to which it belongs. Thus we define the following:

$a_{mc's}(lkpc) =$

$$\begin{cases} 1, & \text{if in plan } l \text{ machine } m \text{ in cell } c' \text{ is assigned to operation } s \text{ for all } (kpc) \\ 0, & \text{otherwise} \end{cases}$$

$h_{cc'}(k) =$ cost of moving part type k from cell c to cell c'.

While computing the inter-cell cost it is assumed from the GT point-of-view that a part, whenever it visits a cell other than the cell it is assigned to, it is always returned to the assigned cell for storage. It is routed to other cells whenever the required machine is free. Thus, an estimate of inter-cell movement will be an overestimate in comparison to the actual inter-cell moves. However, since the exact sequence is not considered at this stage, and we wish to avoid inter-cell moves as much as possible, this can be an acceptable approximation. The cost of assigning a machine to different cells need not be the same. Factors such as equipping the cell with foundation, wiring, accessories etc., influence this cost. The model below includes C_{mc} which is the cost of assigning machine of type m to cell c.

Simultaneous grouping model

Minimize

$$\sum_{mc} C_{mc} Z_{mc} + \sum_{kpcl} \left(\sum_{c'ms} a_{mc's}(lkpc) c_{ms}(kp) \right) X(lkpc) + \sum_{kpcl} \left(\sum_{c'ms} a_{mc's}(lkpc) h_{cc'}(k) \right) X(lkpc)$$

subject to:

$$\sum_{cpl} X(lkpc) \geqslant d_k, \quad \forall k \tag{7.8}$$

$$\sum_{kpl} \left(\sum_{cs} a_{mc's}(lkpc) t_{ms}(kp) \right) X(lkpc) \leqslant b_m Z_{mc'} \quad \forall m, c' \tag{7.9}$$

$$\sum_{m} Z_{mc} \leqslant \text{Max}_c, \quad \forall c \tag{7.10}$$

$$Z_{mc} \text{ have non-negative integer values, } \forall\, m, c;\ X(lkpc) \geqslant 0, \forall l, k, p \tag{7.11}$$

These constraints are similar to constraints 7.4–7.7. As mentioned earlier, the above relaxed model can be solved efficiently using column generation. The solution procedure is discussed next.

Solution methodology

Consider a large-scale mixed integer programming model. The total number of production plans $L(kpc)$ for each (kpc) combination can be extremely large, and a complete enumeration is impractical. The implicitly-known columns correspond to continuous variables $X(lkpc)$. Moreover, the integer restrictions on Z_{mc} have to be taken care of. If the integrality restriction on Z_{mc} is removed, we have a large-scale linear program. Revised simplex can be used to solve this problem. Since the number of variables is large, for subsequent iterations a column generation scheme is used to select the implicitly known columns in the model. Once the relaxed linear program is solved, a branch-and-bound scheme on Z_{mc} will provide an optimal solution. Each node in the branch-and-bound tree represents a solution to an augmented continuous problem with additional constraints on the integer variables. These additional constraints are incorporated without increasing the size of the problem by the bounded variables procedure. The procedure stops when an integer solution is obtained and all the nodes in the branch-and-bound tree are fathomed.

Column generation scheme

The approach considered here is a method of generating the desired column at each iteration of the simplex method. Column generation schemes for solving large-scale linear programs with implicitly-known columns have been considered in the literature. The method of generating the column in each case depends on the structure of the problem. In Gilmore and Gomory (1961) column generation was equivalent to solving a knapsack problem. In both Chandrasekaran, Aneja and Nair (1984) and Ribeiro, Minoux and Penna (1989), though the contexts are different, the column generation was similar and involved solving a sequence of assignment problems. Due to the special structure of the above model, column generation can be achieved much more efficiently by solving a sequence of semi-assignment problems. The solution to the semi-assignment problem is obtained by a 'greedy' procedure which involves sorting a sequence of n numbers. The scheme is presented below.

At any general iteration, define the simplex multipliers corresponding to 7.8 and 7.9 as $\pi_k(k=1,2,\dots K)$ and $u_{mc'}(m=1,2,\dots M;\ c'=1,2,\dots C)$, respectively. The implicitly known columns in this case correspond to continuous variables $X(lkpc)$. Since these variables do not appear in constraints 7.10, the corresponding simplex multipliers $v_c(c=1,2,\dots C)$ do not appear here. Now the pricing scheme for determining the entering variable, if any, is to look for any variable $X\ (lkpc)$ such that the reduced cost $(C(lkpc)=C(lkpc)-Z(lkpc))$ associated with its entry is negative.

Therefore, for a given part k, process plan p assigned to cell c:

$$\sum_{c'ms} a_{mc's}(lkpc)c_{ms}(kp) + \sum_{c'ms} a_{mc's}(lkpc)h_{co'}(k) < \pi_k$$

$$+ \sum_{c'ms} a_{mc's}(lkpc)t_{ms}(kp)u_{mc'}$$

or

$$\sum_{c'ms} a_{mc's}(lkpc)\{c_{ms}(kp) + h_{cc'}(k) + (-u_{mc'})t_{ms}(kp)\} < \pi_k$$

Defining

$$cc_{mc's}(kpc) = \{c_{ms}(kp) + h_{cc'}(k) + (-u_{mc'})t_{ms}(kp)\} \tag{7.12}$$

gives

$$\sum_{c'ms} a_{mc's}(lkpc)cc_{mc's}(kpc) < \pi_k \quad \forall\, k,p,c \tag{7.13}$$

Thus, to generate a column which satisfies equation (7.12) for a fixed k, p, c, consider the following semi-assignment problem of assigning machines to operations.

Define 0-1 variables $a_{mc's}$ as

$$a_{mc's} = \begin{cases} 1, & \text{if machine } m \text{ in cell } c' \text{ is assigned to perform operation } s, \ \forall\, m,s \\ 0 & \text{otherwise} \end{cases}$$

Let $cc_{mc's}$ be the cost of assigning machine m in cell c' to perform operation s. The following semi-assignment problem provides the desired entering column:

Minimize

$$Z = \sum_{c'ms} cc_{mc's}a_{mc's}$$

subject to :

$$\sum_{c'm} a_{mc's} = 1, \quad \forall\, s \tag{7.14}$$

$$a_{mc's} = 0 \text{ or } 1, \quad \forall\, c', m, s \tag{7.15}$$

The optimal solution to this problem is obtained by the following simple 'greedy' procedure. Let

$$m_s = \underset{mc'}{\text{Min}}\ cc_{mc's}, \quad \forall\, s \tag{7.16}$$

The optimal assignment is given by

$$a_{mc's} = \begin{cases} 1, & \text{if } m = m_s \\ 0, & \text{otherwise} \end{cases} \tag{7.17}$$

Let Z^0 be the cost associated with this production plan. If $Z^0 < \pi_k$, then enter the following $(K + MC + C)$ column into the basis: $[e_k, u_c, y_c]$, where e_k is a unit K-vector with 1 at the kth place,

$$u_{c'} = \left[\left(\sum_s a_{lc's} \cdot t_{1s} \sum_s a_{2c's} \cdot t_{2s} \ldots\ldots\ldots, \sum_s aM_{c's} \cdot tM_s,\right)\right]$$

and y_c is a vector with C '0' values. Thus, it can be seen that to determine a column for every operation s, we look at the machines m in all the cells c' and pick the best machine. This is equivalent to looking at mc numbers. This has to be done for all s operations. Thus the computational complexity of the column generation scheme is $O(mcs)$.

If for every (kpc) combination the optimal assignment $Z^0 > \pi_k$, check if any slack, surplus or Z_{mc} variables can enter. These columns are explicity known. If none of these columns can enter, the optimal solution to the relaxed linear program is obtained. If the Z_{mc} variables are integers, the optimal solution to the mixed integer program has been obtained. The complete algorithm is given below.

Algorithm

Step 1. Initial basis will consist of all slack and artificial variables.
Step 2. Choose a part k, process plan p, cell c and the assignment cost matrix as given in equation 7.12.

Step 3. Find the minimal cost assignment such that

$$C^1 = \sum_{c'ms} a_{mc's}(lkpc)\, cc_{mc's}(kpc) < \pi_k, \quad \text{for all } \pi_k \geqslant 0 \text{ and } u_{mg} \leqslant 0$$

If $C^1 > \pi_k$ for all (kpc) combination go to step 6.
Step 4. Enter the new column and update the basis.
Step 5. Check for the surplus and slack variables to enter.

If $\pi_k < 0$, introduce the surplus variable corresponding to park k
$u_{mc'} > 0$, introduce the slack corresponding to machine m in cell c'
If any of them can enter go to step 4, else go to step 6.

Step 6. Check if any Z_{mc} column can enter. If yes then go to step 4, else go to step 7.
Step 7. If Z_{mc} values are integers then stop, else branch-and-bound on Z_{mc}. Add the additional constraints and go to step 1.

Example 7.3

For the purpose of the exposition of the column generation scheme, consider the following information as given:

$C = 2$; $\text{Max}_1 = 4$; $\text{Max}_2 = 2$; $C_{11} = C_{12} = C_{42} = 200$; $C_{21} = C_{31} = C_{22} = 250$; $C_{41} = C_{42} = 350$;

Table 7.6 Production cost and time data for parts (a blank indicate that the operation cannot be performed on the machine)

Cost = time		$m=1$	$m=2$	$m=3$	$m=4$
$k=1$	$s=1$	3		6	
	$s=2$	7		2	
$k=2$	$s=1$		8		3
	$s=2$		2		5
$k=3$	$s=1$	2			8
	$s=2$	9		5	
	$s=3$				3
$k=4$	$s=1$		6	10	
	$s=2$			4	8
	$s=3$		8		5
$k=5$	$s=1$	5		7	
	$s=2$	7		1	
$k=6$	$s=1$		4		1
	$s=2$		2		9

$h_{cc'}(1) = h_{cc'}(2) = 3$; $h_{cc'}(3) = h_{cc'}(5) = h_{cc'}(6) = 2$; $h_{cc'}(4) = 1$, for all $c \neq c'$;
$d_1 = d_3 = d_5 = 20$; $d_2 = d_4 = d_6 = 10$.

Within a cell the material handling cost is taken to be zero. The production cost and time data for all six part types are given in Table 7.6.

This problem has 16 constraints and 8 integer variables. The number of explicit columns in the model is 28 columns corresponding to the production plans and 8 columns corresponding to the machine variables. All the columns corresponding to the production plans need not be explicity listed, instead they will be generated by solving semi-assignment problems. The procedure for generating the columns is explained next. The method begins with all artificial and slack variables in the basis. The initial basic variables column is: the right-hand side column is $[20, 10, 20, 10, 20, 10, 0, 0, 0, 0, 0, 0, 0, 0, 4, 2,]$ and the dual variables are $[M, M, M, M, M, M, 0, 0, 0, 0, 0, 0, 0, 0, 0, 0,]$. M in this context is a very large number. We can take any part, say $k = 1, p = 1, c = 1$ and find the assignment cost $cc_{mc's}(11)$, which is given by Table 7.7. For each operation, the machine with minimum cost is picked up; in this case, machine 1 in cell 1 for operation 1 and machine 3 in cell 1 for operation 2 (the material handling cost of machine 3 was included for operations performed in cell 2, because it was assumed that the part is allocated to cell 1). Thus we have a plan with a cost of 5. This plan does not require any material handling cost because both operations are performed in cell 1. Since this cost is less than M (a large value) it qualifies to enter the basis. The plan column entering the basis is $P^1[1,0,0,0,0,0,3,0,2,0,0,0,0,0,0,0,]$. The basis and the inverse are updated by the usual simplex rules.

Table 7.7 Assignment costs

$cc_{mc's}$ (11)		$s=1$	$s=2$
$c'=1$	$m=1$	3	7
	$m=3$	6	2
$c'=2$	$m=1$	6	10
	$m=3$	9	5

Table 7.8 Assignment of operations to machines in the optimal plans selected

Cost = time		$c=1$				$c=2$			
		$m=1$	$m=2$	$m=3$	$m=4$	$m=1$	$m=2$	$m=3$	$m=4$
$k=1$	s=1	3							
	s=2			2					
$k=2$	s=1								3
	s=2						2		
$k=3$	s=1	2							
	s=2			5					3
	s=3								
$k=4$	s=1						6		
	s=2			4					
	s=3								5
$k=5$	s=1	5							
	s=2			1					
$k=6$	s=1								1
	s=2						2		

(Source: Rajamani, Singh and Aneja (1990); reproduced with permission from Taylor and Francis)

The optimal solution to the problem identifies that we require two machines of each type 1 and 3 in cell 1 and one of each type 2 and 4 in cell 2. The assignment of operations to machines in the plan selected are shown in Table 7.8. Since the number of machines of each type are already integers, we do not have to perform a branch-and-bound on integer variables.

7.4 CELL DESIGN WITH RELOCATION CONSIDERATIONS

Companies that are currently looking towards converting to cellular manufacturing would like to use existing machines rather than purchase new machines during cell design. No additional investment is incurred if existing machines are sufficient to meet the demand for products. Also, with introduction of new parts and changed demands, new part families and machine groups have to be identified. During such redesign, if only existing machines are used the machines in each cell are known. While allocating parts to these cells, material handling capacity might pose a severe constraint. One possible way to minimize the inter-cell movement is to relocate machines (Rajamani, 1990; Rajamani and Szwarc, 1994). If the existing capacity is exceeded we need to know if relocation should be accompanied or substituted by a higher degree of investment on new machines. This will not only enable the company to increase the capacity of the plant to meet the new demand, but also to update its machines to current technology. Gupta and Seifoddini (1990) concluded that one-third of US companies undergo major dislocation of production facilities every two years. A major dislocation in the study was defined as a physical rearrangement of two-thirds or more of the facilities.

The model presented in this section identifies part families and machine groups such that the total relocation expense, of machines as well as the additional cost of material handling, of operating and of new machines, is minimized. Physical limitations such as an upper-bound on cell size, the available machines of each type, machine capacity and material handling capacity are imposed in the model.

Simultaneous grouping model

Minimize

$$\sum_{mcc'} C_{mcc'} Z_{mcc'} + \sum_{mc} C_{mc} Z_{mc} + \sum_{kpcl} \left(\sum_{c'ms} a_{mc's}(lkpc) \cdot c_{ms}(kp) \right) X(lkpc) + \sum_{kpcl} \left(\sum_{c'ms} a_{mc's}(lkpc) \cdot h_{cc'}(k) \right) X(lkpc)$$

subject to:

$$\sum_{cpl} X(lkpc) \geqslant d_k, \quad \forall\, k \tag{7.17}$$

$$\sum_{kpl} \left(\sum_{cs} a_{mc's}(lkpc) \cdot t_{ms}(kp) \right) X(lkpc) \leqslant b_m \left(N_{mc'} + \sum_{c} Z_{mcc'} - \sum_{c} Z_{mc'c} + Z_{mc'} \right), \quad \forall\, m,c' \tag{7.18}$$

$$\sum_{kpcl} \left(\sum_{c'} d_{cc'} \sum_{ms} a_{mc's}(lkpc) \right) X(lkpc) \leqslant D \tag{7.19}$$

$$\sum_{m} N_{mc} + \sum_{mc} Z_{mcc'} - \sum_{mc} Z_{mc'c} + \sum_{m} Z_{mc'} \leqslant \text{Max}_{c'}, \quad \forall c \tag{7.20}$$

$$\sum_{c} Z_{mc'c} - \sum_{c} Z_{mcc'} \leqslant N_{mc'} \quad \forall m,c' \tag{7.21}$$

$$Z_{mcc'}, Z_{mc} \text{ have non-negative integer values, } \forall m,c,c'; X(lkpc) \geqslant 0, \forall l,k,p,c \tag{7.22}$$

where $C_{mcc'}$ is the cost of relocating one machine of type m from cell c to c', $d_{cc'}$ is the distance between cell c and c', N_{mc} is the number of machines of type m in cell c and $Z_{mcc'}$ is the number of machines of type m moved from cell c to cell c'.

Constraints 7.17 force the demand for parts to be met. Constraints 7.18 ensure sufficient capacity is available on machines to process the parts. An upper limit on the material handling capacity is imposed by constraints 7.19. The maximum number of machines that can be in each cell is imposed by constraints 7.20. Constraints 7.21 ensure that the machines relocated to other cells do not exceed the number available in that cell. The integer restrictions are imposed by constraints 7.22.

7.5 CELL DESIGN CONSIDERING OPERATIONAL VARIABLES

Implementing GT results in a well organized cell shop. The literature available is simply not able to determine whether GT is responsible for this benefit or if an improved job shop will give a similar performance. Some researchers (Flynn and Jacobs, 1986; Morris and Tersine, 1990) have studied the performance of GT cells formed by part–machine matrix considerations, and compared them with traditional shops using simulation techniques. The performance of cells thus formed indicates that cellular systems perform more poorly in terms of work-in-process inventory, average job waiting time and job flow times than the improved job shops. However, they have superior performance in terms of average move times and setup times. The main reason for the poor performance is that current cell design procedures do not consider operational aspects during cell formation.

To illustrate the impact of operational variables, this section considers cell formation in flow line manufacturing situations similar to those involved in repetitive manufacturing. The parts require the same set of machines in the same order. This situation arises in a number of chemical and process industries. Typical examples include manufacture of paints, detergents etc. The setups incurred during changeovers are usually sequence-dependent. For example, in the manufacture of paints the equipment must be cleaned when there is a change from one color to another. The thoroughness of the cleaning is heavily dependent on the color being removed and the color for which the machine is being prepared.

In a sequence-dependent manufacturing environment, where demand for parts is repetitive and the production requirements are similar, the sequence in which to produce the parts can be selected such that the total cost and time spent on setup is minimized. The sequence thus determined may give a schedule in which parts finish early or late compared to their due dates. In addition, there may be part waiting between machines, or machine idle time. Alternatively, there could be a separate line for producing each part, which would avoid cost and time lost due to sequence dependence. Also, the inventory could be reduced by synchronizing the production rate of cells with the demand rates. However, the investment cost in this case is high.

Clearly, investment options between these two extremes are also available. For example, late finishing of parts can be avoided by increasing the capacity of bottleneck stages or by re-sequencing them after adding a new cell. The sequence of parts also affects the work-in-progress and utilization of machines. Achieving minimum inventory and minimum machine idle time are conflicting objectives as reduction in one often leads to an increase in the other. Depending on the scenario, the appropriate parameters should be considered and weighted accordingly. This section presents the model proposed by Rajamani, Singh and Aneja (1992a) which considers only the trade-off between investment and sequence-dependent setup costs. For a mathematical model which considers the trade-offs between investment and operational costs (sequence-dependent setup, machine idle time, part inventory, part early and late finish) refer to Adil, Rajamani and Strong (1993).

Notation

$c = 1, \ldots\ldots C$ cells
$j = 1, \ldots\ldots c$ positions
$k, l = 1, \ldots\ldots K$ parts
$m = 1, \ldots\ldots M$ machines
t_{km} time for machine m to perform operation on part k
S_{kl} setup cost incurred if part k is followed by part l
T_{kl} setup time incurred if part k is followed by part l
Z_{mc} number of machines of type m in cell c

$$X_{kj}^{c} = \begin{cases} 1, & \text{if part } k \text{ is assigned to position } j \text{ in cell } c \\ 0, & \text{otherwise} \end{cases}$$

$$Y_{klj}^{c} = \begin{cases} 1, & \text{if part } k \text{ is assigned to position } j-1 \\ 0, & \text{and part } l \text{ to position } j \text{ in cell } c \geqslant 2 \text{ otherwise and in } c = 1 \end{cases}$$

The maximum number of cells which can be formed is equal to the number of cells. To minimize the number of variables 0–1, we define distinct points in each cell to capture the sequence of parts in each cell. Thus, the kth cell will contain k points. For example, if three parts are considered, the following points are defined:

cell 1 *
cell 2 **
cell 3 ***

The above six points are sufficient to capture all arrangement possibilities for the three part types. Only three of these points will be assigned and the rest will be unassigned. With the above definition we will have only 18 variables 0–1. A typical definition of a variable to capture the sequence dependence would be $X_{kl}^{c}=1$ if part k precedes part l in cell c; 0 otherwise. This definition for the three-part problem will require the definition of 27 variables 0–1. The mathematical model is given as

Minimize

$$\sum_{mc} C_m Z_{mc} + \sum_{cjkl} S_{kl} Y_{klj}^{c}$$

subject to:

$$\sum_{cj} X_{kj}^{c} = 1, \quad \forall k \tag{7.23}$$

$$\sum_{k} X_{kj}^{c} \leqslant 1, \quad \forall c, j=1 \tag{7.24}$$

$$\sum_{k} X_{k,j+1}^{c} \leqslant \sum_{k} X_{kj}^{c}, \quad \forall c, j \tag{7.25}$$

$$Y_{klj}^{c} \geqslant X_{k,j-1}^{c} + X_{kj}^{c} - 1, \quad \forall i, l, c, j \tag{7.26}$$

$$\sum_{jk} d_k t_{km} X_{kj}^{c} \sum_{jkl} T_{kl} Y_{klj}^{c} \leqslant b_m Z_{mc}, \quad \forall m, c \tag{7.27}$$

$$Y_{klj}^{c} \geqslant 0;\ X_{kj}^{c} = 0/1;\ Z_{mc} \text{ is a general integer} \tag{7.28}$$

The objective of the model is to minimize the sum of discounted cost of machines assigned to cells and the setup costs incurred due to the sequence dependence of parts in each cell. Constraints 7.23 guarantee that each part is produced in one of the cells. Constraints 7.24 ensure that the first position in each cell can be assigned to at most one part. Constraints 7.25 ensure that the $(j+1)$th position in a cell can be assigned only if the jth position is assigned. These constraints also ensure that not more than one part is assigned to all other positions except 1. The sequence in which a part is assigned to a cell is uniquely determined through constraints 7.26. Constraints 7.27 ensure that the required machine capacity is available in the cells to meet the demand.

In many practical situations, the number of parts produced in repetitive manufacturing is not great. However, the problem size becomes large with an increase in part types. However, in such situations, the above model can be effectively used by aggregating the part types with similar setup into fewer families.

Example 7.4

A soft-drinks company mixes and bottles five different product flavors. The standard cost and times for changing the production facility, which consists of three machines, from one flavor to another are shown in Table 7.9. The information on process times, demand for each flavor, the capacity of the production facility and the discounted cost of machines is given in Table 7.10. The company wishes to determine the number of production lines to be purchased and the sequence in which the flavors should be mixed in each line.

The model identifies three cells to be formed, where products 2,5 and 4 (in that sequence) are identified in the same cell, and products 1 and 3 are allocated to independent cells. The additional investment on new machines is less than the savings obtained on setup by identifying parts 1 and 3 in the same cell. Details of the number of machines in each cell are given in Table 7.11.

7.6 RELATED DEVELOPMENTS

Chakravarthy and Shtub (1984) presented an approach to generate an efficient layout of machines in groups and also to establish production

Table 7.9 Setup-dependent costs and time

Costs($)	*Cola*	*Grape*	*Orange*	*Beer*	*Lime*
Cola	0	18	10	10	10
Grape	2	0	4	3	2
Orange	5	18	0	8	10
Beer	6	17	7	0	8
Lime	4	12	5	4	0
Time (min)					
Cola	0	16	17	4	24
Grape	20	0	22	3	24
Orange	17	20	0	3	22
Beer	34	26	30	0	32
Lime	21	17	20	3	0

Table 7.10 Process times demand capacity and discounted cost of machines

Machine type	*1*	*2*	*3*	*4*	*5*	*Capacity on machine (min per shift)*	*Discounted cost of machine per shift*
$m=1$	10	2	7	7	5	100	15
$m=2$	8	3	9	3	2	100	10
$m=3$	7	4	6	1	8	100	20
Demand per shift	10	20	10	30	20		

Table 7.11 Optimum number of cells, parts and number of machines in each cell

	Cell 1	*Cell 2*	*Cell 3*
Parts	1	3	2,5,4
Machines:			
$m=1$	1	1	4
$m=2$	1	1	3
$m=3$	1	1	3

lot sizes of parts to match the layout. Co and Araar (1988) presented a three-stage procedure for configuring machines into manufacturing cells and assigning the cells to process a specific set of jobs. Choobineh (1988) presented a two-stage procedure where in the first stage part families are identified by considering the manufacturing sequences, in the second stage, an integer programming model was proposed to specify the type and number of machines required for the objective of minimizing investment and operational costs. Askin and Chiu (1990) presented a mathematical model to consider the costs of inventory, machine depreciation, machine setup and material handling. The model is divided into two sub-problems to facilitate decomposition. A heuristic graph partitioning procedure was proposed for each sub-problem. Balasubramanian and Pannerselvam (1993) developed an algorithm based on a covering technique to determine the economic number of manufacturing cells and the arrangement of machines within each cell. The design process considers the sequence of part visits and minimizes the handling cost, machine idle time and overtime.

Irani, Cavalier and Cohen (1993) introduced an approach which integrates machine grouping and layout design, not considering part family formation. The concepts of hybrid and cellular layout and virtual manufacturing cells are discussed. They showed that the combination of overlapping GT cells, functional layout and handling reduces the need for machine duplication among cells.

Shafer and Rogers (1991) presented a goal programming model for the cell formation problem. The model considers a number of design objectives such as reducing setup times, minimizing inter-cell movement, minimizing investment and maintaining an acceptable level of machine utilization. Only one process route is assumed for each part and the impact of sequence on setup is considered in the model. For efficient solution, they presented a heuristic solution by partitioning the goal programming model into two sub-problems and solving them in successive stages.

Frazier, Gaither and Olson (1990) provided a procedure for dealing with multiple objectives. Heragu and Kakuturi (1993) presented a three-stage approach. They integrated the machine grouping and layout problem, in which the objective was not only to identify machine cells and corresponding part families, but also to determine a near-optimal layout of machines within each cell and the cells themselves. Material flow considerations and alternate process plans can be considered while determining the machine groups. Operational aspects such as the impact of refixturing were considered by Damodaran, Lashkari and Singh (1992). Sankaran and Kasilingam (1993) developed a mathematical model to capture the exact sequence of parts and considered the effect of cell size on the intra-cell handling cost; the intra-cell handling cost increases as a step function with an increase in number of machines assigned to a cell. A heuristic procedure was also presented, which can be used in some special situations. For the selection of a subset of parts and machines for cellularization, see Rajamani, Singh and Aneja (1992b).

7.7 SUMMARY

Cells are formed using new and often automated machines and material handling systems. A judicious selection of processes and machines is necessary for cell formation.

With the introduction of new parts and changed demands, new part families and machine groups have to be identified. The redesign of such systems warrants consideration of practical issues such as the relocation expense of existing machines, investment on new machines etc. The creation of exclusive cells with no inter-cell movement is a common goal for cell formation. However, often it is not economical to achieve cell independence. Material handling is an important aspect to be considered in this situation. In fact, new technology and faster deterioration of certain machines could render the previous allocation of parts and machines undesirable. Thus, there is also a need to determine if the old machines must be replaced with new or technologically updated machines. This chapter provided a mathematical framework to address many of these issues.

Table 7.12 Time and cost information for operations on compatible machines for different process plans

		$k=1$		$k=2$		$k=3$			$k=4$	
		$p=1$	$p=2$	$p=1$	$p=2$	$p=1$	$p=2$	$p=3$	$p=1$	$p=2$
$s=1$	$m=1$	3,5		4,3		2,2		1,8	2,1	7,9
	$m=3$	2,7		3,4		2,2		2,9	1,2	9,8
$s=2$	$m=2$	3,5	8,9	8,7	3,3	3,3	2,1	9,5	3,2	8,9
	$m=3$	4,3	9,7	7,7	3,2	4,4	4,2	10,3	4,2	9,10
$s=3$	$m=1$		8,8	9,10	5,6		7,11	4,7	5,3	
	$m=2$		7,7	8,9	6,6		8,8	9,5	2,6	

Cell formation as defined in this chapter, in addition to identifying part families and machine groups, specifies the plans selected for each part, the quantity to be produced through the selected plans, the machine type to perform each operation in the plans, the total number of machines required, the machines to be relocated, and the parts and machines to be selected for cellularization considering demand, time, material handling and resource constraints. Some pertinent objectives considered were the minimization of investment, operating cost, machine relocation cost, material handling cost, and the maximization of output. Consideration of physical limitations such as the upper bound on cell size, machine capacity, material handling capacity etc. was also incorporated in the cell design process.

PROBLEMS

7.1 Illustrate by an example how alternate process plans can lead to better cell formation.

7.2 Four different part types of known demand ($d_1 = d_2 = d_3 = d_4 = 50$) are manufactured each with 2, 2, 3 and 2 process plans as given in Table 7.12. Each operation in a plan can be performed on alternate machines. Three types of machines of known capacity ($b_1 = b_2 = b_3 = 500$) and discounted cost ($C_1 = 1250$; $C_2 = 500$; $C_3 = 1500$) are available. The time and cost information for performing an operation on compatible machines for each process plan is also given in Table 7.12. Solve the sequential model assuming parts 1 and 2 belong to the first part family and parts 3 and 4 are identified as the second part family. Solve the simultaneous model for $C = 2$. Solve the model for few combinations of Max_c. Compare the two results obtained for the given situation.

7.3 Use the column generation approach to solve Q 7.2.

Table 7.13 Setup-dependent costs and time

Costs (s)	*Red*	*White*	*Orange*	*Yellow*
Red	0	9	5	5
White	2	0	4	3
Orange	5	9	0	8
Yellow	6	10	7	0
Time (min)				
Red	0	8	9	4
White	20	0	12	3
Orange	9	10	0	3
Yellow	24	19	20	0

Table 7.14 Process times, demand, capacity and the discounted cost of machines

Machine type	*1*	*2*	*3*	*4*	*Capacity on machine (min per shift)*	*Discounted cost of machine per shift*
$m = 1$	2	10	7	7	100	15
$m = 2$	3	7	9	3	100	10
$m = 3$	4	7	1	6	100	20
Demand per shift	30	10	10	20		

7.4 A paint company mixes and bottles four different colors. The standard cost and times for changing the production facility, which consists of three machines, from one color to another are given in Table 7.13. Information on process times, the demand for each color, the production capacity, and the discounted cost of machines are also known (Table 7.14).

The company wishes to determine the number of production lines to be purchased and the sequence in which the colors should be mixed in each line such that the total cost is minimized.

REFERENCES

Adil, G. K., Rajamani, D. and Strong D. (1993) A mathematical model for cell formation considering investment and operational costs. *European Journal of Operational Research*, **69**(3), 330–41.

Askin, R. G. and Chiu, K. S. (1990) A graph partitioning procedure for machine assignment and cell formation in group technology. *International Journal of Production Research*, **28**(8), 1555–72.

Balasubramanian, K N. and Pannerselvam, R. (1993) Covering technique based algorithm for machine grouping to form manufacturing cells. *International Journal of Production Research*, **31**(6), 1479–504.

Burbridge, J. L. (1992) Change to group technology: a process organization is obsolete. *International Journal of Production Research*, **30**(5), 1209–19.
Chandrasekaran, R., Aneja, Y. P. and Nair, K. P. K (1984) Production planning in assembly line systems. *Management Science*, **30**(6), 713–19.
Chakravarthy, A. K. and Shtub, A. (1984) An integrated layout for group technology within process inventory costs. *International Journal of Production Research*, **22**(3), 431–42.
Choobineh, F. (1988) A framework for the design of cellular manufacturing systems. *International Journal of Production Research*, **26**(7), 1161–72.
Co, H. C. and Araar, A. (1988) Configuring cellular manufacturing systems. *International Journal of Production Research*, **26**(9), 1511–22.
Damodaran, V., Lashkari, R. S. and Singh, N. (1992) A production planning model for cellular manufacturing systems with refixturing considerations. *International Journal of Production Research*, **30**(7), 1603–15.
Flynn, B. B. and Jacobs, F. R. (1986) A simulation comparison of group technology with traditional job shop manufacturing. *International Journal of Production Research*, **24**(5), 1171–92.
Frazier, G. V., Gaither, N. and Olson, D. (1990) A procedure for dealing with multiple objectives in cell formation. *Journal of Operations Management*, **9**(4), 465–80.
Gilmore, P. C. and Gomory, R. E. (1961) A linear programming approach to cutting stock problem. *Operations Research*, **9**, 849–59.
Gupta, T. and Seifoddini, H. (1990) Production data based similarity coefficient for machine-component grouping decisions in the design of a cellular manufacturing system. *International Journal of Production Research*, **28**(7), 1247–69.
Heragu, S. S. and Kakuturi, S. R. (1993) Grouping and placement of machine cells. Rensselaer Poly. Inst., Troy, NY. Working paper.
Irani. S. A., Cavalier, T. M. and Cohen. P. H. (1993) Virtual manufacturing cells: exploiting layout design and intercell flows for the machine sharing problem. *International Journal of Production Research*, **31**(4), 791–810.
Morris, J. S. and Tersine, R. J. (1990) A simulation analysis of factors influencing the attractiveness of group technology cellular layouts. *Management Science*, **36**(12), 1567–78.
Rajamani, D. (1990) Design of cellular manufacturing systems. Univ. Windsor, Ontario, Canada. Doctoral dissertation.
Rajamani, D., Singh, N. and Aneja, Y. P. (1990) Integrated design of cellular manufacturing systems in the presence of alternate process plans. *International Journal of Production Research*, **28**(8), 1541–54.
Rajamani D., Singh, N. and Aneja, Y. P. (1992a) A model for cell formation in manufacturing systems with sequence dependence. *International Journal of Production Research*, **30**(6), 1227–35.
Rajamani, D., Singh, N. and Aneja, Y. P. (1992b) Selection of parts and machines for cellularization: a mathematical programming approach. *European Journal of Operational Research*, **62**(1), 47–54.
Rajamani, D., Singh, N. and Aneja, Y. P. (1993) Design of cellular manufacturing systems. Univ. Manitoba, Canada. Working paper.
Rajamani, D. and Szwarc, D. (1994) A mathematical model for multiple machine replacement with material handling and relocation consideration. *Engineering Optimization*, **22**(2), 213–29.
Ribeiro, C. C., Minoux, M. and Penna, M. C. (1989) An optimal column generation with ranking algorithm for very large set partitioning problems in traffic assignment. *European Journal of Operational Research*, **41**, 232–9.

Sankaran, S. and Kasilingam, R. G. (1993) On cell size and machine requirements planning in group technology systems. *European Journal of Operational Research*, **69**(3), 373–83.

Shafer, S. M. and Rogers, D. F. (1991) A goal programming approach to the cell formation problem. *Journal of Operations Management*, **10**(1), 28–43.

Wemmerlov, U. and Hyer, N. L. (1989) Cellular manufacturing in the US industry: a survey of current practices. *International Journal of Production Research*, **27**(8), 1287–304.

CHAPTER EIGHT

Layout planning in cellular manufacturing

Almost everyone has some experience of layout planning in terms of arranging facilities (furniture, appliances, and so forth) in the house or office. Recall how many times you have changed the arrangement of furniture in your study room or how many times you have changed the location of the television in your house. Every time you do it, knowingly or unknowingly you do some layout planning. If layout planning is so common and trivial, why have researchers bothered about this for so long and why has it been a subject of so many books and papers? The reason is that it is a common decision in a variety of situations but not a trivial one in all types of situation, as the cost of undoing it will differ significantly. For example, furniture in the house may be rearranged for a diminutive expenditure. On the other hand, rearrangement of machines in a manufacturing system could cost a fortune.

Decisions on the specific location and design of facilities for a given space based on some long-term objectives are crucial. This is part of layout planning and it has long-term implications for any manufacturing organization. A facility layout plan should emerge from the overall strategic plan of the organization. Factors to be considered for layout planning may be broadly classified as internal and external. Most of the internal factors have a two-way relationship with layout decisions. For example, the volume of workflow may be a major decision variable for layout but once the layout is final, the volume itself will depend on the layout type. External factors such as market demand for the product will definitely affect decisions on the layout, but not vice-versa.

Layout planning is a science as well as an art. Although it relies heavily on systematic techniques and mathematical modeling, for effective layout planning one has to go beyond the limitations of these principles and guidelines. To develop a good layout, an in-depth understanding of the system is essential so that one can improvise on the available scientific methods and tools. This chapter provides a discussion of the types of layouts and modeling approaches used for layout planning with the emphasis on layout planning for cellular manufacturing systems.

8.1 TYPES OF LAYOUT FOR MANUFACTURING SYSTEMS

There are four basic types of layout used for manufacturing systems:

- fixed layout
- product layout
- process layout
- group / cell layout

Product, process and group / cell layouts can be distinguished based on system characteristics such as production volume and product variety relationships (Suer and Ortega, 1994; Steudel and Desruelle, 1992) as shown in Fig. 8.1. Accordingly, a particular type of layout or a combination of layouts can be selected to meet the internal and external requirements of the production system. Each of these layouts is briefly discussed below.

Fixed position layout

The concept of fixed position layout differs from other types of layout. For example, production equipment moves to the product manufacturing site in the case of fixed position layout, as shown in Fig. 8.2. In contrast, products move to the manufacturing site in the case of other layouts. Fixed position layout is used for products which cannot move or are very heavy, such as building construction, ship building, aircraft

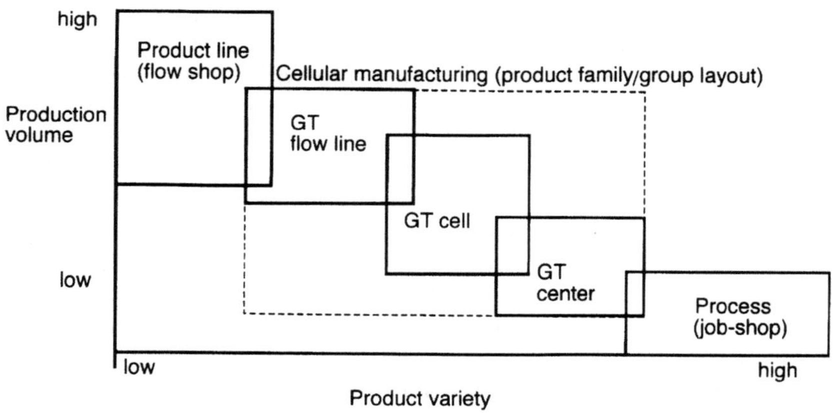

Fig. 8.1 Product volume and variety relationships with different manufacturing systems.

manufacturing and so forth. Project-type organizations are most associated with fixed position layouts. This type of system may result in duplication of facilities. Production control and coordination is complex as, in general, are the space requirements and in-process inventory.

This type of layout is flexible and can accommodate changes in design, volume and product mix. In fact, each product produced may be unique, buildings for example. Material movement is reduced considerably. However, personnel and equipment movements are higher. Skill requirements on the part of personnel are high. Task identification is more difficult for workers in this type of system. In most cases, a team approach to tasks is used, which provides job enrichment opportunities.

Product layout

Product layouts are associated with high-volume production and low product variety. In product layouts, facilities are arranged in the sequence of operations required, as shown in Fig. 8.3. Depending on the type of product, these layouts may be flow-type or line-type. Flow-type layouts are related to continuous production such as in the chemical industry. Line-type layouts, however, are associated with discrete manufacturing such as in the automotive industry.

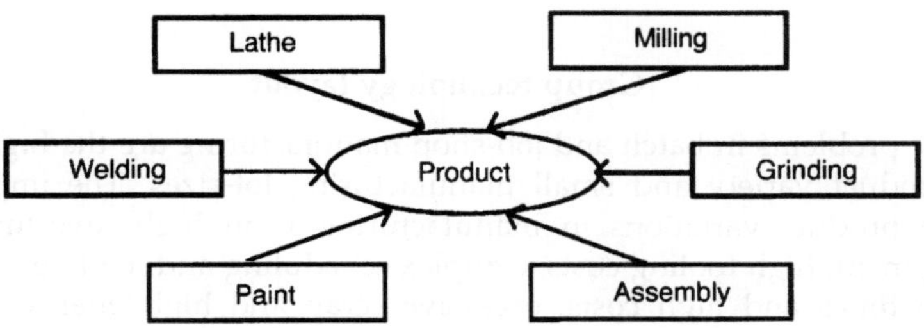

Fig. 8.2 Fixed position layout.

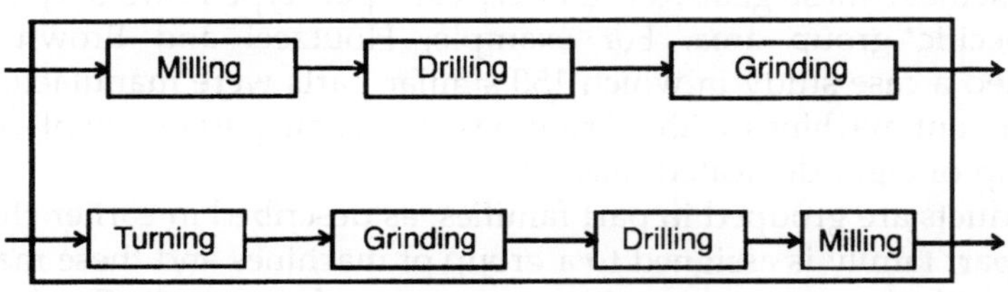

Fig. 8.3 Product layout.

Product-type layouts require special-purpose equipment, and investment in this equipment is high. If the product changes, it may require changes in the layout, which may be costly. This is one of the reasons why flexibility is very low in such layouts. The labor skill requirement is low as most of the tasks are simple and repetitive. Sometimes this can result in motivational problems. Material flow is smooth, simple and logical. Production control is therefore simpler for product layouts. Accordingly, material handling requirements are reduced; manufacturing lead times are shorter and inventories lower. However, the system requires highly reliable equipment since failure at one workstation may cause the stoppage of the whole line.

Process layout

Process layouts are associated with low-volume production and high product variety, such as in batch or job-shop manufacturing systems. Process layouts are achieved by grouping like processes together, as shown in Fig. 8.4. Equipment used in this type of layout is mostly general-purpose. The labor skill requirement is high because a variety of jobs have to be handled. Material flow is not as smooth as in product layouts. Flexibility is higher but efficiency is lower. Investment in equipment is lower and utilization is higher. The production system is more complicated and operational-level decisions such as scheduling and loading are important and difficult. The inventory level is higher and manufacturing lead times are longer.

Group technology layout

Major problems in batch and job-shop manufacturing are the high level of product variety and small manufacturing lot-sizes. The impact of these product variations in manufacturing is in high investment in equipment, high tooling costs, complex scheduling and loading, lengthy setup times and high costs, excessive scrap and high quality control costs. Adoption of GT concepts enables small batch production to gain economic advantages similar to mass production while retaining the flexibility of the job-shop. Organizing facilities according to a GT layout helps achieve these goals, since in GT each part type flows only through its specific group area. For example, Houtzeel and Brown (1984) reported a case study in which 150 similar parts were manufactured on 51 different machines with 87 routings. The same parts were placed into a group of eight dedicated machines.

Products are grouped in part families, as described in earlier chapters. Each part family is assigned to a group of machines and these machines along with the material handling equipment form a cell. The layout of these facilities is known as a group technology, or cell, layout. This

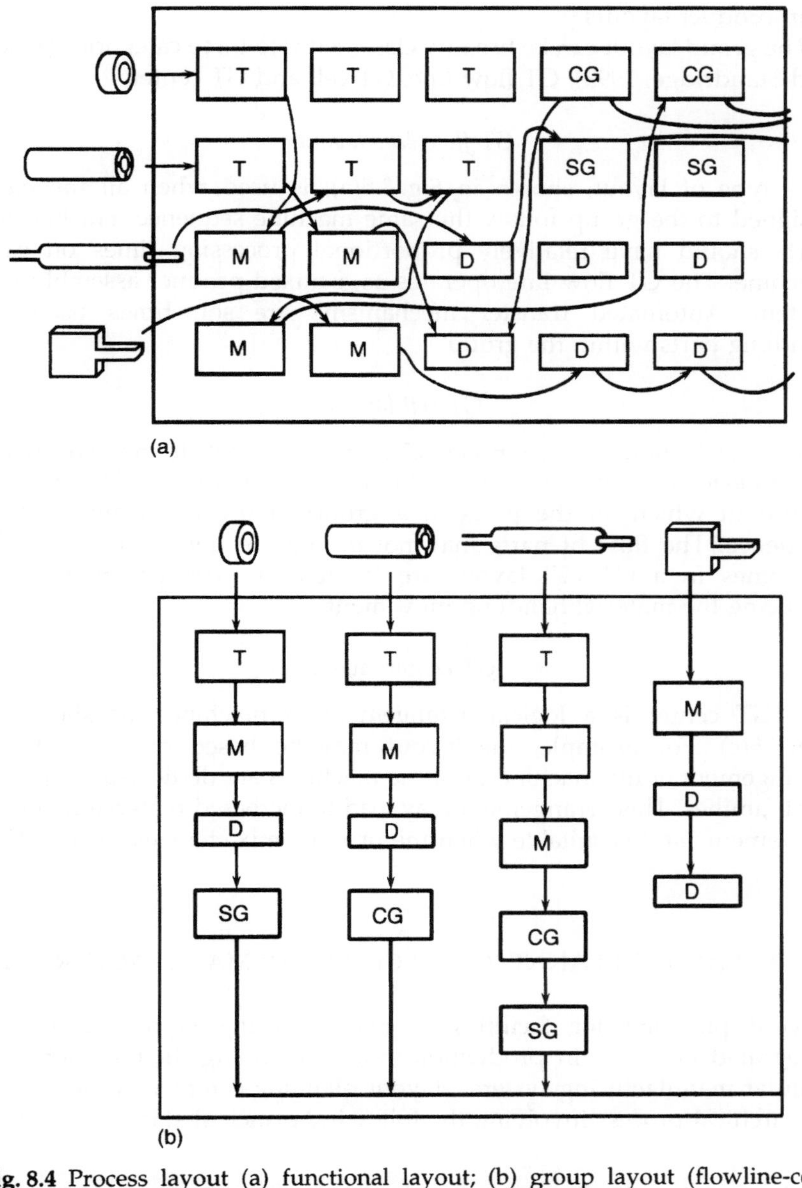

Fig. 8.4 Process layout (a) functional layout; (b) group layout (flowline-cell). (Published with permission from the Decision Sciences Institute.)

layout will be discussed in detail in the following sections. Group layout has some of the benefits of product and process layouts. Equipment is generally computer-controlled and can handle a variety of tasks and sequences, which gives these systems a very high flexibility. GT layouts provide higher efficiency than process layouts and are more flexible than product layouts.

The group layout can be broadly classified into three categories (Askin and Standridge, 1993): GT flow line, GT cell and GT center.

GT flow line layout

This type of layout, shown in Fig. 8.5(a), is used when all the parts assigned to the group follow the same machine sequence. Further, the parts should have relatively proportional processing times on each machine. The GT flow line operates as a mixed-product assembly line system. Automated transfer mechanisms are sometimes used for handling parts within the group.

GT cell layout

The GT cell layout, shown in Fig. 8.5(b), permits parts to move from any machine to any other machine. This contrasts with the GT flow line layout in which all the parts in a group follow the same machine sequence. The flow of parts may not be unidirectional. However, the machines in a GT cell layout are located in close promixity, thus reducing the material handling movement.

GT center layout

The GT center is a logical arrangement of machines, as shown in Fig. 8.5(c). For example, the layout may be based on a functional arrangement of the machines, but the machines are dedicated to specific part families. This arrangement may lead to increased material handling movements and is suitable when the product-mix changes frequently.

8.2 LAYOUT PLANNING FOR CELLULAR MANUFACTURING

Layout planning for facilities in manufacturing is one of several integrated elements in production system planning. In the context of cellular manufacturing systems, layout planning can be considered as a hierarchical process involving the following principal stages:

1. determining families of parts based on part design and process similarities;
2. assigning part families to groups of machines (cells);
3. rationalization of part families and workloads;
4. selecting the type of cell layout;
5. laying out machines and auxiliary facilities in cells.

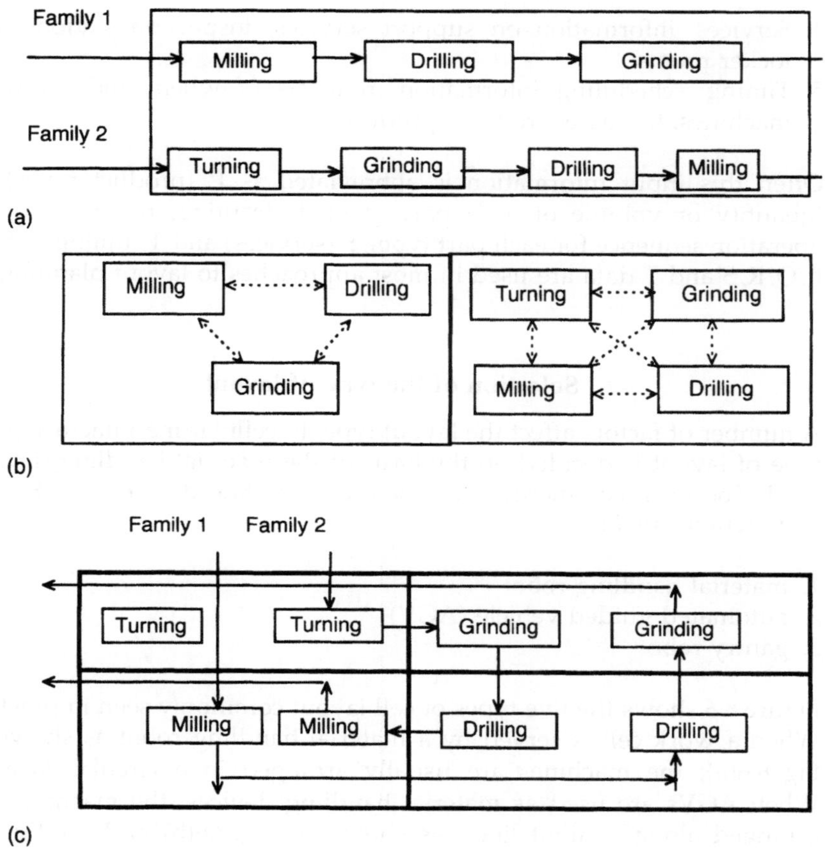

Fig. 8.5 GT layouts: (a) GT flow line; (b) GT cell; (c) GT center.

The first three stages have been discussed in detail in previous chapters, so here we will concentrate on the last two stages. However, before a layout plan can be developed, the following information is needed.

1. Characteristics of products and materials: types of products and materials, their sizes and shapes. Product variety is a major factor in designing facilities for economical production.
2. Product quantities: it is important to know the present and future quantities of each product. The product variety and quantity relationships dictate the type of layout to a large extent, as shown Fig. 8.1.
3. Process routing: information on sequences in which products are processed.

4. Services: information on support services, inspection stations and locker rooms.
5. Timing: scheduling information in terms of when, and on what machines, the parts are to be produced.

Often this input information is abbreviated as P (product types), Q (quantity or volume of each part type), R (routing, referring to the operation sequence for each part type), S (services) and T (timing). These P, Q, R, S and T data are used in most approaches to layout planning.

Selection of the type of layout

A number of factors affect the layout type. In cellular manufacturing the type of layout is decided on the basis of the material handling devices used. The most commonly used material handling devices are (Kusiak and Heragu, 1987):

1. material handling robot
2. automated guided vehicle (AGV)
3. gantry robot

Figure 8.6 shows the five types of cell layout commonly seen in practice. When a work cell is served by a material handling robot as shown in Fig. 8.6(a), the machines are usually arranged in a circular fashion. When AGVs are used as material handling devices, the machines are arranged along straight lines, as shown in Fig. 8.6(b) and (c). Such a straight line arrangement is necessary because an AGV serves most efficiently while traveling along straight paths. When space is a limiting factor, gantry robots are used to transfer parts among machines, as shown in Fig. 8.6(d). In such cases the geometry of the layout is not important. The limitations here are of a different nature:

- the size of the machines
- the working envelope of the gantry robot
- access of the robot arm to the machines.

The layout shown in Fig. 8.6(e) is often used in flexible manufacturing cells. Material handling is accomplished by the conveyor system which allows only unidirectional flow of parts around the loop. A secondary material handling system is also provided at each workstation which permits the flow of parts without any obstruction. Many other variations to the loop layout are possible, for example, a ladder layout contains rungs on which workstations are located; an open-field layout consists of loops, ladders and sidings.

Layout of machines and auxiliary facilities

The laying out of machines and auxiliary facilities is a strategic decision, and it is a process with several stages:

1. defining the problem of layout planning;
2. selecting the solution methodology;
3. generating alternative plans;
4. modifying and selecting one or two plans;
5. deciding the implementation procedure;
6. readjustment and revision based on the implementation and initial problems.

A large number of objectives are involved in layout planning (Hales, 1984):

- effective movement of materials and personnel
- effective utilization of space
- adaptability to unforeseen changes
- easy expansion
- control of noise
- safety
- easy supervision and control
- good appearance
- security
- low cost

There are a variety of constraints for any real-life layout planning problem. The most common planning constraints (Hales, 1984) include:

- one or more fixed activities
- activities which must be separated
- architectural limitations
- material handling limitations
- utility limitations
- organizational restrictions
- budget
- time

Many of these objectives are conflicting. The goal of attaining all the layout objectives and at the same time satisfying the relevant constraints is a challenging task and may not be achievable because of conflicting objectives. Therefore, a compromise solution is often sought. Once proper objectives and constraints are defined and input from other production planning stages are available, what is then needed is a solution methodology appropriate for the layout planning problem. Such approaches can be classified into two broad categories:

- traditional approaches
- mathematical programming-based approaches

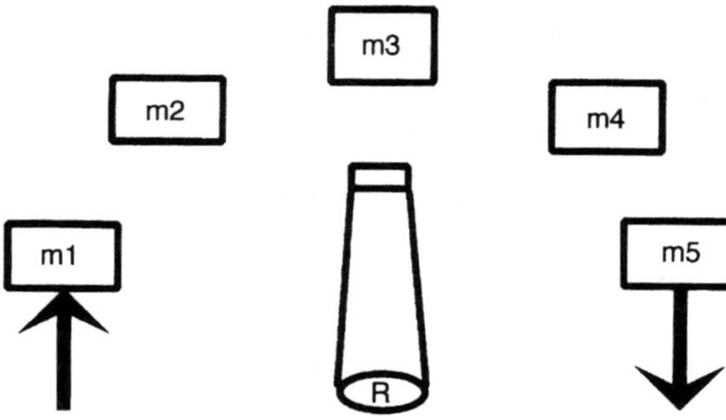
m3
m2
m4
m1
m5
R

(a)

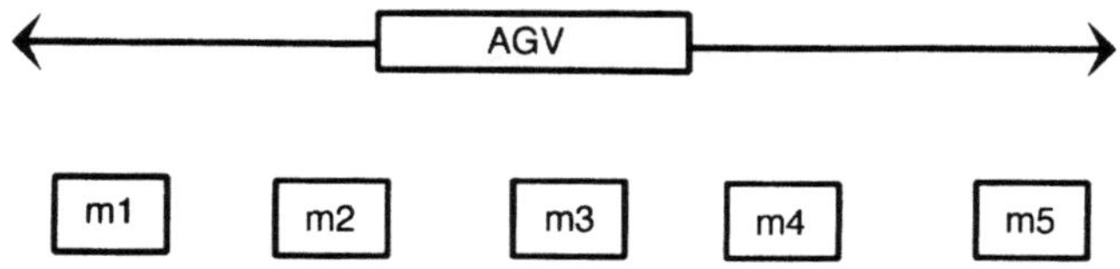
AGV
m1
m2
m3
m4
m5

(b)

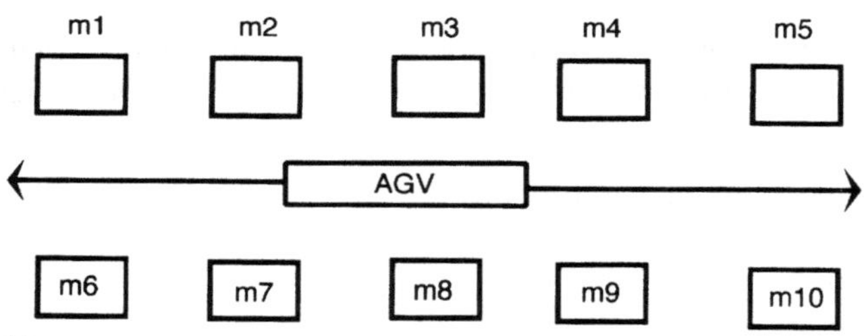
m1
m2
m3
m4
m5
AGV
m6
m7
m8
m9
m10

(c)

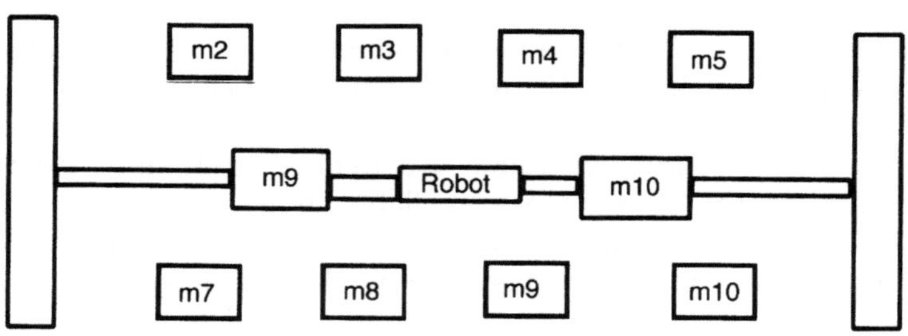
m2
m3
m4
m5
m9
Robot
m10
m7
m8
m9
m10

(d)

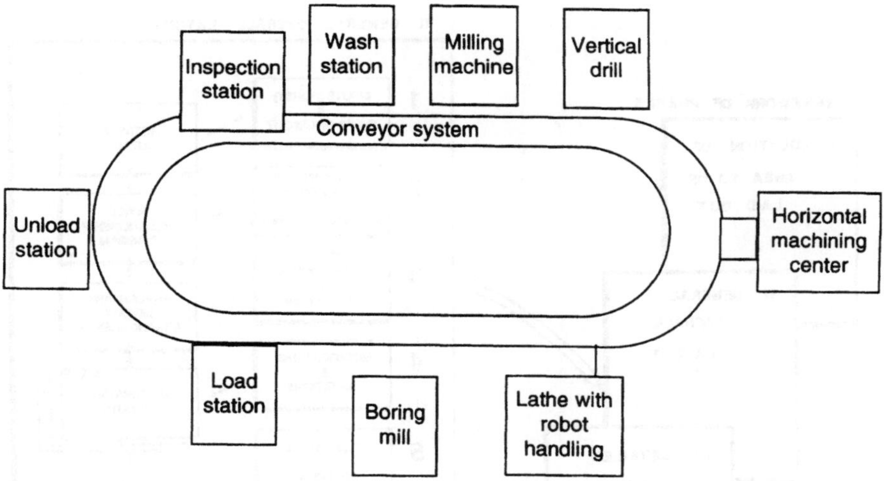

(e)

Fig. 8.6 Layout of material handling devices: (a) single cell robot; (b) single row layout; (c) double row layout; (d) cell layout with gantry robot; (e) flexible cell layout. (Source: Singh N., *Systems Approach to Computer-integrated Design and Manufacturing*, ©1996, Reprinted by permission of John Wiley & Sons, Inc. New York.)

Traditional approaches

Prominent among traditional approaches are:

- Apple's approach (Apple, 1977)
- Reed's approach (Reed, 1961)
- the systematic layout planning (SLP) approach

There are many commonalities in these approaches. SLP (Muther, 1973) is one of the most commonly used traditional approaches and can be described as a four-stage process:

phase I: determining the area location
phase II: establishing overall layout
phase III: developing detailed layout
phase IV: installing the best layout

The first phase identifies the area within an existing building or in a new building where the facilities are to be laid out. A multi-step interactive

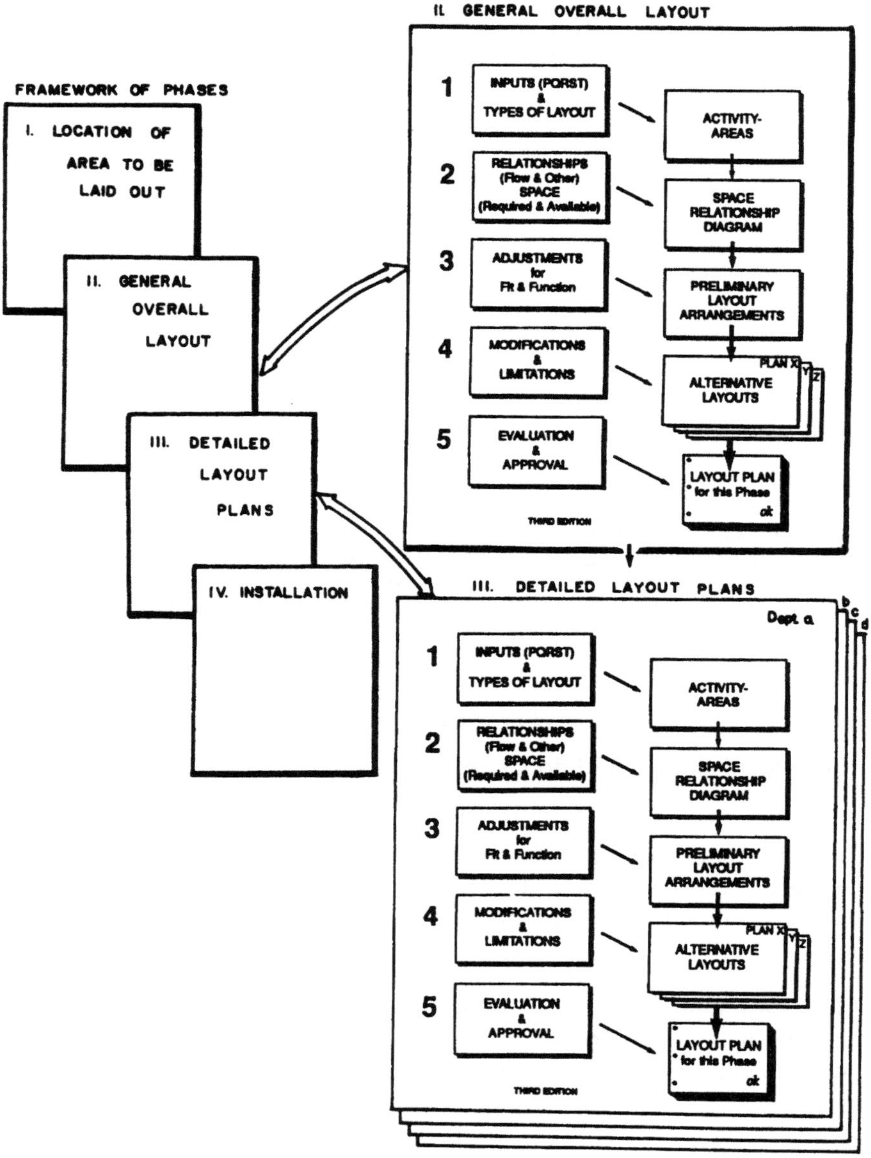
FRAMEWORK OF PHASES
I. LOCATION OF AREA TO BE LAID OUT
II. GENERAL OVERALL LAYOUT
III. DETAILED LAYOUT PLANS
IV. INSTALLATION
II. GENERAL OVERALL LAYOUT
1
INPUTS (PQRST) & TYPES OF LAYOUT
ACTIVITY-AREAS
2
RELATIONSHIPS (Flow & Other) SPACE (Required & Available)
SPACE RELATIONSHIP DIAGRAM
3
ADJUSTMENTS for Fit & Function
PRELIMINARY LAYOUT ARRANGEMENTS
4
MODIFICATIONS & LIMITATIONS
ALTERNATIVE LAYOUTS
PLAN X Y Z
5
EVALUATION & APPROVAL
LAYOUT PLAN for this Phase ok
THIRD EDITION
III. DETAILED LAYOUT PLANS
Dept a
b
c
d
1
INPUTS (PQRST) & TYPES OF LAYOUT
ACTIVITY-AREAS
2
RELATIONSHIPS (Flow & Other) SPACE (Required & Available)
SPACE RELATIONSHIP DIAGRAM
3
ADJUSTMENTS for Fit & Function
PRELIMINARY LAYOUT ARRANGEMENTS
4
MODIFICATIONS & LIMITATIONS
ALTERNATIVE LAYOUTS
PLAN X Y Z
5
EVALUATION & APPROVAL
LAYOUT PLAN for this Phase ok
THIRD EDITION

Fig. 8.7 (a)

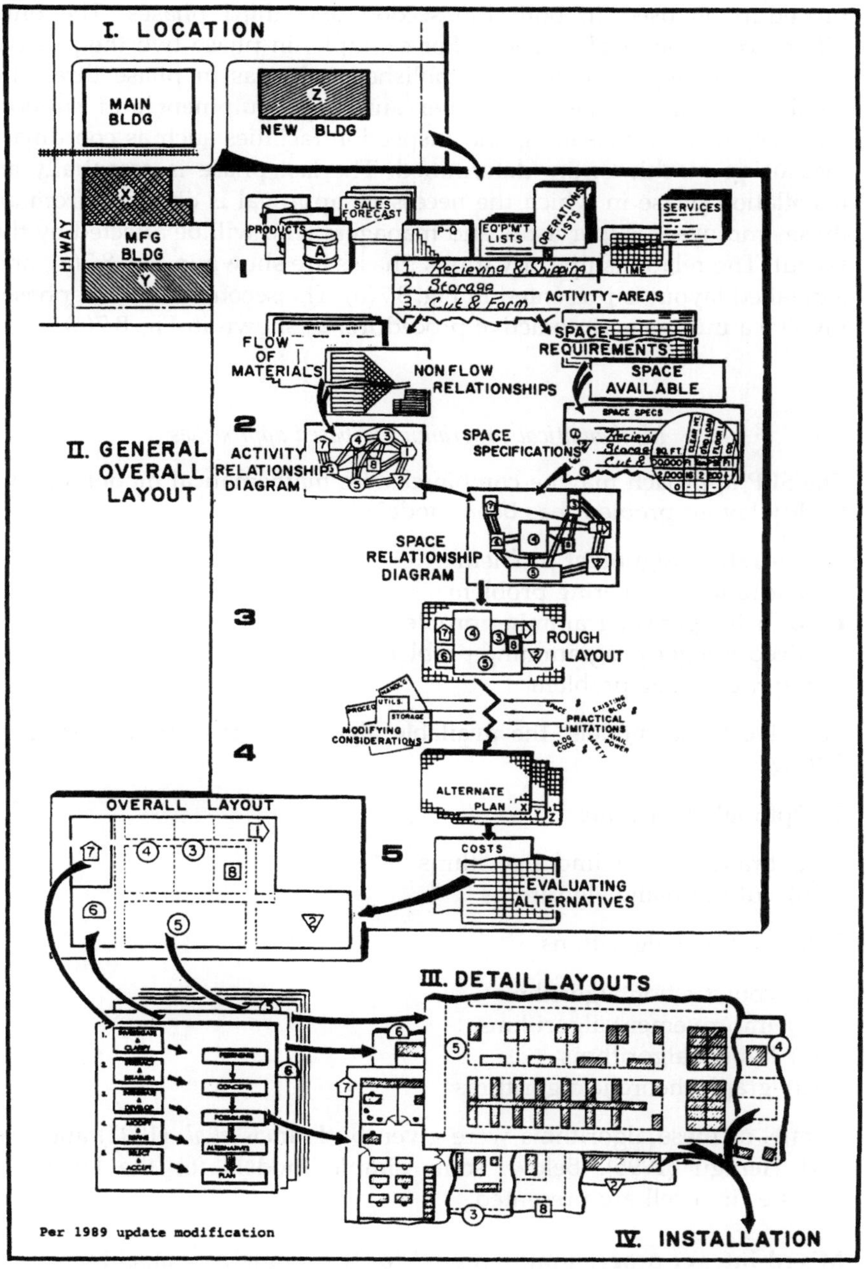

Fig. 8.7 Systematic layout planning : (a) phases; (b) detailed layout. (Source: © 1989, Richard Muther and Associates. Reproduced with permission.)

procedure is used in both the second and third phases. The only difference is the level of detail. For example, in phase two the relative positions of departments are established, whereas in phase three the detailed layout of machines, other auxiliary equipment and support services, including cleaning and inspection facilities such as coordinate measuring machines, are determined. The last phase is essentially an installation phase in which the necessary approval is obtained from all those employees, supervisors and managers who will be affected by the layout. The relationships of various phases are shown in Fig. 8.7(a); and a detailed layout is presented in Fig. 8.7(b). The second and third phases involve a multi-step interactive procedure, as shown in Fig. 8.7(a).

Mathematical programming-based approaches

The SLP approach may be combined with mathematical modeling. The facility layout problem has been modeled as a

- quadratic assignment problem
- quadratic set covering problem
- linear integer programming problem
- mixed integer programming problem
- graph theoretic problem

To solve these models, the available algorithms may be classified as follows:

1. Optimal algorithms
   (a) branch-and-bound algorithms
   (b) cutting plane algorithms
2. Suboptimal algorithms
   (a) construction algorithms
   (b) improvement algorithms
   (c) hybrid algorithms
   (d) graph theoretic algorithms

Details of these algorithms were given by Francis *et al.* (1992) and Das and Heragu (1995). Below, some simple models for the layout of facilities in a cell are presented.

Mixed integer programming model for the single row machine layout problem in a cell

This section presents an analytical model for the single row machine layout problem based on the work of Neghabat (1974) and Kusiak (1990). The machines are arranged in a straight line. The objective is to determine the non-overlapping optimal sequence of the machines such that the total cost of making the required trips between machines is minimized. Consider the following notation:

Notation

c_{ij} material handling cost per unit distance between all pairs of machines for all $(i,j), i \neq j$
f_{ij} frequency of trips between all pairs of machines (frequency matrix) for all (i,j), $i \neq j$
l_i length of ith machine; d_{ij} is the clearance between machines i and j.
n number of machines
x_j distance of jth machine from the vertical reference line, as shown in Fig. 8.8.

Minimize
$$f = \sum_{i=1}^{n-1} \sum_{j=i+1}^{n} c_{ij} f_{ij} |x_i - x_j|,$$
$$|x_i - x_j| \geq (1/2)(l_i + l_j) + d_{ij},$$
$$\forall (i,j),\ i = 1,2,\ldots,n-1, j = i+1, i+2, \ldots, n \quad (8.1)$$
$$x_i \geq 0 \quad \forall\ i = 1,2,\ldots,n \quad (8.2)$$

The objective function represents the total cost of trips between the machines. Constraints 8.1 ensure that there is no overlap between the machines. The non-negativity constraint is given by 8.2. The absolute terms in the model can be easily transformed, resulting in an equivalent linear integer programming model. Such a model is given next. Although the model can be solved by standard linear programming packages, a simple heuristic algorithm is provided in the following section. The following additional variables are defined to transform the absolute terms in the formulation:

$$x_{ij}^{+} = \begin{cases} (x_i - x_j), & \text{if } (x_i - x_j) > 0 \\ 0, & \text{if } (x_i - x_j) \leq 0 \end{cases}$$

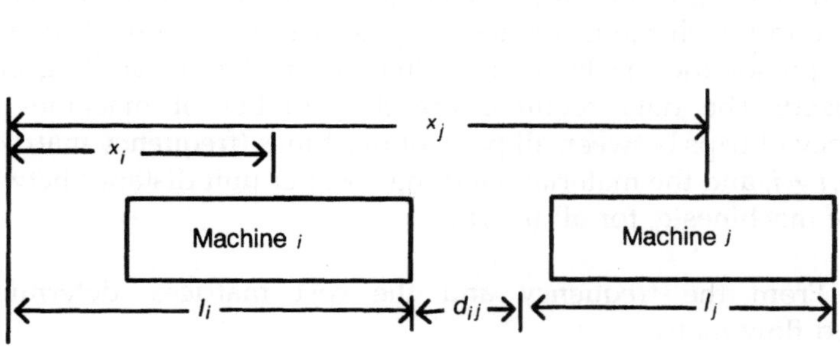

Fig. 8.8 Machine location relative to reference line.

$$x_{ij}^- = \begin{cases} -(x_i - x_j) & \text{if } (x_i - x_j) < 0 \\ 0, & \text{if } (x_i - x_j) \geqslant 0 \end{cases}$$

$$z_{ij} = \begin{cases} 1, & \text{if } x_i < x_j \\ 0, & \text{if } x_i \geqslant x_j \end{cases}$$

Accordingly, the equivalent model is

$$f = \min \sum_{i=1}^{n-1} \sum_{j=i+1}^{n} C_{ij} f_{ij} (x_{ij}^+ + x_{ij}^-) \tag{8.3}$$

subject to:

$$x_i - x_j + M z_{ij} \geqslant 1/2(l_i + l_j) + d_{ij}, \quad i = 1, \ldots, n-1, \quad j = i+1, \ldots, n \tag{8.4}$$

$$-(x_i + x_j) + M(1 - z_{ij}) \geqslant 1/2(l_i + l_j) + d_{ij}, \quad i = 1, \ldots, n-1, \quad j = i+1, \ldots, n \tag{8.5}$$

$$x_{ij}^+ - x_{ij}^- = x_i - x_j, \quad i = 1, \ldots, n-1, \; j = i+1, \ldots, n \tag{8.6}$$

$$x_{ij}^-, x_{ij}^- \geqslant 0, \quad i = 1, \ldots, n-1, j = i+1, \ldots, n \tag{8.7}$$

$$x_i \geqslant 0, \quad i = 1, \ldots, n \tag{8.8}$$

$$z_{ij} = 0, 1, \quad i = 1, \ldots, n-1, j = i+1, \ldots, n \tag{8.9}$$

where M is a large positive number.

As z_{ij} is a 0,1 variable, only one of constraints 8.4 and 8.5 hold and ensure that no two machines coincide. Constraints 8.7 and 8.8 ensure non-negativity. This mixed integer programming model may be solved using any standard algorithm. For an optimum solution, the available algorithms require a lot of computer memory and time. There are many simple and efficient heuristic algorithms available for solving such problems. One such algorithm proposed by Heragu and Kusiak (1988) is described below.

Heuristic algorithm for circular and linear single row machine layout
This heuristic algorithm due to Heragu and Kusiak (1988) provides the sequence in which the machines are placed in the layout. The objective is to sequence the machines such that the material handling effort is minimized. The data required are the number of machines n, the frequency of trips between all pairs of machines (frequency matrix) f_{ij} for all $(i, j), i \neq j$, and the material handling cost per unit distance between all pairs of machines c_{ij} for all $(i, j), i \neq j$.

Step 1. From the frequency and the cost matrices, determine the adjusted flow matrix:

$$\bar{F} = [\bar{f}_{ij}] = [f_{ij} \cdot c_{ij}]$$

Step 2. Determine $\bar{f}_{i'j'} = \max\ [\bar{f}_{ij},$ for all i and $j]$. Obtain the partial solution by connecting i' and j'. Set $\bar{f}_{i'j'} = \bar{f}_{j'i'} = -\infty$

Step 3. Determine

$$\bar{f}_{p'q'} = \max\ [\bar{f}_{i'k}, \bar{f}_{j'l};\quad k = 1, 2, \ldots, n; l = 1, 2, \ldots, n]$$

(a) Connect q' to p' and q' to the partial solution.
(b) Delete row p' and column p' from $[\bar{f}_{ij}]$.
(c) If $p' = i'$, set $i' = q'$; otherwise set $j' = q'$.

Step 4. Repeat step 3 until all the machines are included in the solution.

Example 8.1

Consider five machines in a flexible manufacturing system which have to be served by an AGV. A linear single row layout is recommended because an AGV is to be used. Data on the frequency of AGV trips, material handling costs per unit distance and the clearance between the machines are given in Fig. 8.9(a)–(c) and Table 8.1. Suggest a suitable layout. (This example is adopted from Singh, 1996. Reproduced with the permission of John Wiley & Sons, Inc., New York.)

(a)

	1	2	3	4	5
1	0	20	70	50	30
2	20	0	10	40	15
3	70	10	0	18	21
4	50	40	18	0	35
5	30	15	21	35	0

(b)

	1	2	3	4	5
1	0	2	7	5	3
2	2	0	1	4	2
3	7	1	0	1	2
4	5	4	1	0	3
5	3	2	2	3	0

(c)

	1	2	3	4	5
1	0	2	1	1	1
2	2	0	1	2	2
3	1	1	0	1	2
4	1	2	1	0	1
5	1	2	2	1	0

Fig. 8.9 (a) Frequency of trips between pairs of machines; (b) cost matrix; (c) clearance matrix.

Table 8.1 Machine dimensions

Machine	*m1*	*m2*	*m3*	*m4*	*m5*
Machine sizes	10 × 10	15 × 15	20 × 30	20 × 20	25 × 15

	1	2	3	4	5
1	0	40	490	250	90
2	40	0	10	160	30
3	490	10	0	18	42
4	250	160	18	0	105
5	90	30	42	105	0

Fig. 8.10 Adjusted flow matrix.

Step 1. Determine the adjusted flow matrix (Fig. 8.10).
Step 2. Include machines 1 and 3 in the partial solution as they are connected.
Step 3. Add machine 4 to the partial solution as it is connected to machine 1. Delete row 1 and column 1.
Step 3. Add machine 2 to the partial solution as it is connected to machine 4. Delete row and column 4 from the matrix.
Step 3. Add machine 5 to the partial solution.
Step 4. Since all the machines are connected, stop. The final sequence is 5, 2, 4, 1, 3. It is obtained by arranging the adjusted flow weights in increasing order while retaining the connectivity of the machines. Accordingly, the final layout considering the clearances between the machines is given in Fig. 8.11.

Model for the multi-row layout problem with machines of unequal area

A nonlinear program is used to model the layout problem in which the machines are of unequal areas. It is assumed that the machines are rectangular in shape; also, the physical orientation of the machines is assumed to be known. The following parameters are defined:

Notation

b_i width of machine i
c_{ij} cost of transporting a unit of material from location i to location j
d^h_{ij} horizontal clearance
d^v_{ij} vertical clearance
f_{ij} flow of material from plant i to plant j
l_i length of machine i

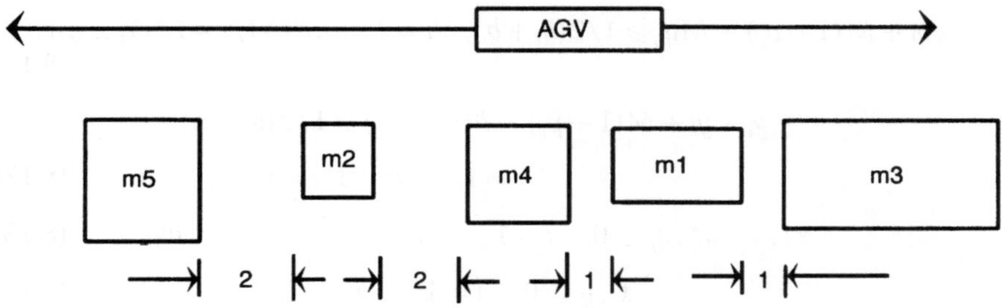

Fig. 8.11 Final layout of the linear single row machine problem.

x_i horizontal distance between machine i and the vertical reference line
y_i vertical distance between machine i and the horizontal reference line

The objective function of this model minimizes the total cost involved in making the required trips between the machines:

Min

$$\sum_{i=1}^{n-1}\sum_{j=i+1}^{n} c_{ij}f_{ij}(|x_i - x_j| + |y_i - y_j|)$$

subject to:

$$|x_i - x_j| + Mz_{ij} \geqslant 1/2(|l_i - l_j|) + d^h_{ij}, \quad i=1,\ldots,n-1,\ j=i+1,\ldots,n \quad (8.10)$$

$$|y_i - y_j| + M(1-z_{ij}) \geqslant 1/2(|b_i - b_j|) + d^v_{ij}, \quad i=1,\ldots,n-1, j=i+1,\ldots,n \quad (8.11)$$

$$z_{ij}(1-z_{ij}) = 0, \quad i=1,\ldots,n-1,\ j=i+1,\ldots,n \quad (8.12)$$

$$x_i, y_i \geqslant 0, \quad i=1,\ldots,n \quad (8.13)$$

Constraints 8.10 and 8.11 ensure that no two machines in the layout overlap. Constraint 8.12 ensures that only one of first and second constraints holds. Constraint 8.13 ensures non-negativity. This model may be transformed into an equivalent linear mixed integer programming model that is shown below.

The objective function of this model minimizes the total cost involved in making the required trips between the machines:

$$\text{Min} \quad \sum_{I=1}^{n-1}\sum_{j=i+1}^{n} c_{ij}f_{ij}(x^+_{ij} + x^-_{ij} + y^+_{ij} + y^-_{ij})$$

subject to:

$$x_i - x_j + M(p_{ij} + q_{ij}) \geqslant 1/2(l_i + l_j), \quad i=1,\ldots,n-1,\ j=i+1,\ldots,n \quad (8.14)$$

$$-x_i + x_j + Mp_{ij} + M(1-q_{ij}) \geqslant 1/2(l_i + l_j), \quad i=1,\ldots,n-1, j=i+1,\ldots,n \quad (8.15)$$

$$y_i - y_j + M(1-p_{ij}) + Mq_{ij} \geqslant 1/2(b_i + b_j), \quad i=1,\ldots,n-1, j=i+1,\ldots,n \tag{8.16}$$

$$-y_i + y_j + M(1-p_{ij}) + M(1-q_{ij}) \geqslant 1/2(b_i + b_j)$$
$$i=1,\ldots,n-1, j=i+1,\ldots,n \tag{8.17}$$

$$x_{ij}^+, x_{ij}^-, y_{ij}^+, y_{ij}^- \geqslant 0, \quad i=1,\ldots,n-1, j=i+1,\ldots,n \tag{8.18}$$

$$x_i, y_i \geqslant 0, \quad i=1,\ldots,n \tag{8.19}$$

$$p_{ij}, q_{ij} = 0,1 \quad i=1,\ldots,n-1, j=i+1,\ldots,n \tag{8.20}$$

The first four constraints of this model ensure that no two machines in the layout overlap: the fifth constraint ensures that only one of the first four constraints holds; the last two constraints ensure non-negativity.

Quadratic assignment model

The multi-row layout problem for cellular manufacturing may be modeled as a quadratic assignment problem (QAP). Koopmans and Beckmann (1957) developed the following QAP model for layout problems.

Notation

a_{ij} net revenue from operating machine i at location j
c_{jl} cost of transporting a unit of material from location j to l
f_{ik} flow of material from machine i to k
n total number of locations

$$x_{ij} = \begin{cases} 1, & \text{if machine } i \text{ is at location } j \\ 0, & \text{otherwise} \end{cases}$$

Assumptions of the model are: a_{ij} includes the difference of total revenue and primary investment, but does not include the transportation cost between machines; f_{ik} is independent of the location of the machines; and c_{jl} is independent of machines and it is cheaper to transport material directly from machine i to machine k than through a third location.

Max

$$f = \sum_{i=1}^{n}\sum_{j=1}^{n} a_{ij}x_{ij} - \sum_{i=1}^{n}\sum_{j=1}^{n}\sum_{k=1}^{n}\sum_{l=1}^{n} f_{ik}c_{jl}x_{ij}x_{kl}$$

subject to:

$$\sum_{j=1}^{n} x_{ij} = 1, \quad i=1,\ldots\ldots\ldots,n \tag{8.21}$$

$$\sum_{i=1}^{n} x_{ij} = 1, \quad j=1,\ldots\ldots\ldots,n \tag{8.22}$$

$$x_{ij} = 0 \text{ or } 1, \quad i = 1, \ldots\ldots, n, \; j = 1, \ldots\ldots, n \tag{8.23}$$

The QAP is NP-complete. The largest problem for which an optimal solution has been found is for 15 facilities (Burkard 1984). The mixed integer non-linear programming model may be extended to model multi-row problems. To solve the QAP, several suboptimal algorithms are available. One such algorithm is CORELAP (Lee and Moore 1967) which belongs to the class of construction algorithms.

All of these mathematical programming models consider only one objective: minimizing material handling cost. Although it is possible to construct a mathematical model with multi-objectives for a layout problem this type of model would be too complex. Therefore, we may take the output of single objective model and modify it based on other objectives and constraints using a systematic layout planning approach. Initially, two or three layouts may be selected to be considered for detailed implementation. Next, based on issues such as the design of aisles, temporary storage, services, lighting, level of noise, psychological and social implications and communication, one layout should be finalized.

Now the details of the implementation for the final selected layout should be decided. As emphasized earlier, layout planning is a strategic decision and involves the commitment of large numbers of resources for the long-term. Therefore, implementation requires careful consideration and planning. Project management techniques such as Program Evaluation and Review Technique (PERT) may be used at this stage.

8.3 DESIGN OF ROBOTIC CELLS

The use of robotic cells is common in industry (Asfahl, 1992). Section 8.2 studied a circular layout in which machines are served by a robot. An important measure of the performance of these cells is the throughput rate, which depends on the sequencing of robot moves as well as the layout of machines and robots. Viswanadham and Narahari (1992) and Asfahl (1992) provided procedures to determine the cycle time for two- and three-machine robotic cells served by a single robot considering only sequential robot moves. However, for a single robot cell with n machines, the number of possible alternative sequences of robot moves is $n!$. To obtain the optimal cycle time, and consequently the best sequence of robot moves, Sethi *et al.* (1992) completely characterized single-robot cells with two and three machines. This section presents a simplified algorithm based on these works for determining the optimal sequence of robot moves to minimize the cycle time in the cases of two- and three-machine robotic cells (adopted from Singh, 1996; reproduced with the permission of John Wiley & Sons, Inc., New York).

This analysis can be used to determine the best cycle times of various cell layouts. For example, three machines can be served by two robots resulting in a number of configurations: one robot serving one or two machines. If the improvements in cycle time are economically justified, an appropriate robotic cell layout and sequence of robot moves for that cell layout can be selected. The following analysis will be needed to arrive at such decisions, or simulation models can be used to help select the best layout.

Sequencing robot moves in a two-machine robotic cell

For the following two-machine robotic cell the two alternative robot sequences are shown in Fig. 8.12(a) and (b). I and O represent the input pickup and output release points respectively.

In alternative 1 (Fig. 8.12 (a)), the robot picks up a part at I, moves to machine m1, loads the part on m1, waits at m1 until the part has been processed, unloads the part from m1, moves to m2, loads the part on m2, waits at m2 until the part has been processed, unloads the part from m2, moves to O, drops the part at O, and moves back to I.

In alternative 2 (Fig. 8.12 (b)), the robot picks up a part, say p1, at I, moves to m1, loads p1 on machine m1, waits at m1 until p1 has been processed, unloads p1 from m1, moves to m2, loads p1 on m2, moves to I, picks up another part p2 at I, moves to m1, loads p2 on m1, moves to m2, if necessary waits at m2 until the earlier part p1 has been processed, unloads p1, moves to O, drops p1 at O, moves to m1, if necessary waits at m1 until the part p2 has been processed, unloads p2, moves to m2, loads p2 on M2, and moves to I.

The cycle time T_1 for alternative 1 is

$$T_1 = \varepsilon + \delta + \varepsilon + 3\delta + \varepsilon + \delta + \varepsilon + a + \varepsilon + \delta + \varepsilon + b = 6\varepsilon + 6\delta + a + b \qquad (8.24)$$

where a and b are the processing times of machines m1 and m2, respectively; ε is the time for each pickup, load, unload and drop operation; and δ is the robot travel time between any pair of adjacent locations.

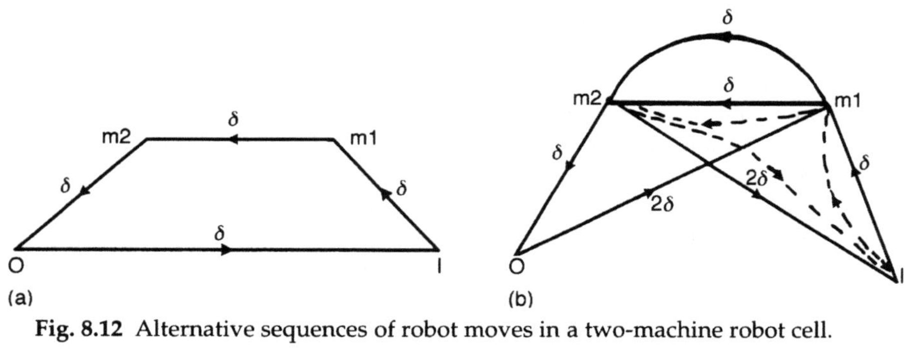

Fig. 8.12 Alternative sequences of robot moves in a two-machine robot cell.

In case of alternative 2, the cycle can begin at any instant. For ease of representation, we begin with the unloading of machine m2. Unload m2, move and leave part at O, move to m1, wait if necessary, otherwise unload part at m1 move to m2 and leave part at m2, move to I and pickup part at I, move to m1 and release, move to m2 and wait if necessary, and pickup part at m2. The cycle time T_2 for alternative 2 is

$$\begin{aligned} T_2 &= \varepsilon + \delta + \varepsilon + 2\delta + w_1 + \varepsilon + \delta + \varepsilon + 2\delta + \varepsilon + \delta + \varepsilon + \delta + w_2 \\ &= 6\varepsilon + 8\delta + w_1 + w_2 \end{aligned} \tag{8.25}$$

where w_1 and w_2 are the robot waiting times at m1 and m2, respectively:

$$w_1 = \max\{0, a - 4\delta - 2\varepsilon - w_2\}$$

$$w_2 = \max\{0, b - 4\delta - 2\varepsilon\}$$

Note that the component $6\varepsilon + 8\delta$ on the right-hand side of equation 8.25 can be split into two components $\alpha = 4\varepsilon + 4\delta$ and $\mu = 2\varepsilon + 4\delta$. Then (8.25) becomes

$$\begin{aligned} T_2 &= \alpha + \mu + w_2 + w_1 \\ &= \alpha + \mu + \max\{0, b - \mu\} + \max\{0, a - \mu - \max\{0, b - \mu\}\} \\ &= \alpha + \max\{\mu, b\} + \max\{0, a - \max\{\mu, b\}\} \\ &= \alpha + \max\{\max\{\mu, b\}, a\} \\ &= \alpha + \max\{\mu, b, a\} = 4\varepsilon + 4\delta + \max\{2\varepsilon + 4\delta, a, b\} \end{aligned} \tag{8.26}$$

where $\alpha + \mu$ represents the total time of the robot activities (pickup, drop off and move times) in a cycle. Then α represents the total time of the robot activities associated with any directed triangle (m2–O–m1 or m1–m2–I in Fig. 8.12(b)) in the cycle, while μ represents the total time of the remaining robot activities.

To determine the optimal cycle time, we must determine the conditions under which one alternative has a minimum cycle time, i.e. one dominates the other. Using equation 8.26 for T_2, consider the following cases:

1. If $\mu \leqslant \max\{a,b\}$, then T_2 is either $\alpha + a$ or $\alpha + b$. In both the cases, by comparing T_2 with T_1, it is found that T_2 is less than T_1.
2. If $\mu > \max\{a,b\}$ and $2\delta \leqslant a + b$, then T_2 is less than T_1.
3. If $\mu > \max\{a,b\}$ *and* $2\delta > a + b$, then T_2 is more than T_1.

These cases can be conveniently represented in algorithmic form.

Algorithm

Step 0. Calculate $\mu = 2\varepsilon + 4\delta$.
Step 1. If $\mu \leqslant \max\{a,b\}$, then T_2 is optimal. Calculate T_2 and stop, otherwise go to step 2.

Step 2. If $\mu > \max\{a,b\}$ and $2\delta \leqslant a+b$, then T_2 is optimal. Calculate T_2 and stop, otherwise go to Step 3.
Step 3. If $\mu > \max\{a,b\}$ and $2\delta > a+b$, then T_1 is optimal. Calculate T_1 and stop.

Example 8.2

Determine the optimal cycle time and corresponding robot sequences in the case of a two-machine robotic cell with the following data (adopted from Singh, 1996; reproduced with the permission of John Wiley & Sons, Inc., New York).

processing time of m1 = 11.00 min
processing time of m2 = 09.00 min
robot gripper pickup = 0.16 min
robot gripper release time = 0.16 min
robot move time between the two machines = 0.24 min.

Step 0. $\mu = 2\varepsilon + 4d$, i.e. $\mu = 2(0.16) + 4(0.24) = 1.28$ min.
Step 1. $\mu \leqslant \max\{a,b\}$, i.e. $1.28 \leqslant \max\{11,9\}$; 1.28 is less than 11, so T_2 is optimal. $T_2 = \alpha + \max\{\mu,a,b\} = [4\varepsilon + 4\delta] + \max\{1.28,11,9\} = [4(0.16) + 4(0.24)] + 11 = 12.6$ min.

The optimal cycle time is 12.6 min and the optimal robot sequence is given by Fig. 8.12 (b).

Sequencing robot moves in a three-machine robotic cell

In this case there are six alternatives, as shown in Fig. 8.13(a)–(f). The cycle time for each alternative can be determined as given below.

The cycle time T_1 for alternative 1 (Fig. 8.13(a)) is

$$T_1 = 8\varepsilon + 8\delta + a + b + c \quad \text{or} \quad T_1 = \alpha + \beta - 4\delta + a + b + c$$

where $\alpha = 4\varepsilon + 4\delta$, $\beta = 4\varepsilon + 8\delta$ and a, b and c are the processing times at machines m1, m2 and m3, respectively (ε and δ have the same meaning as in the case of a two-machine cell).

The cycle time T_2 for alternative 2 (Fig. 8.13 (b)) is

$$T_2 = 8\varepsilon + 12\delta + w_1 + w_2 + w_3$$

where w_1, w_2 and w_3 are the robot waiting times at machines m1, m2 and m3, respectively:

$$w_1 = \max\{0, a - 2\varepsilon - 4\delta - w_2\}$$
$$w_2 = \max\{0, b - 4\varepsilon - 8\delta - w_3\}$$
$$w_3 = \max\{0, c - 2\varepsilon - 4\delta\}$$

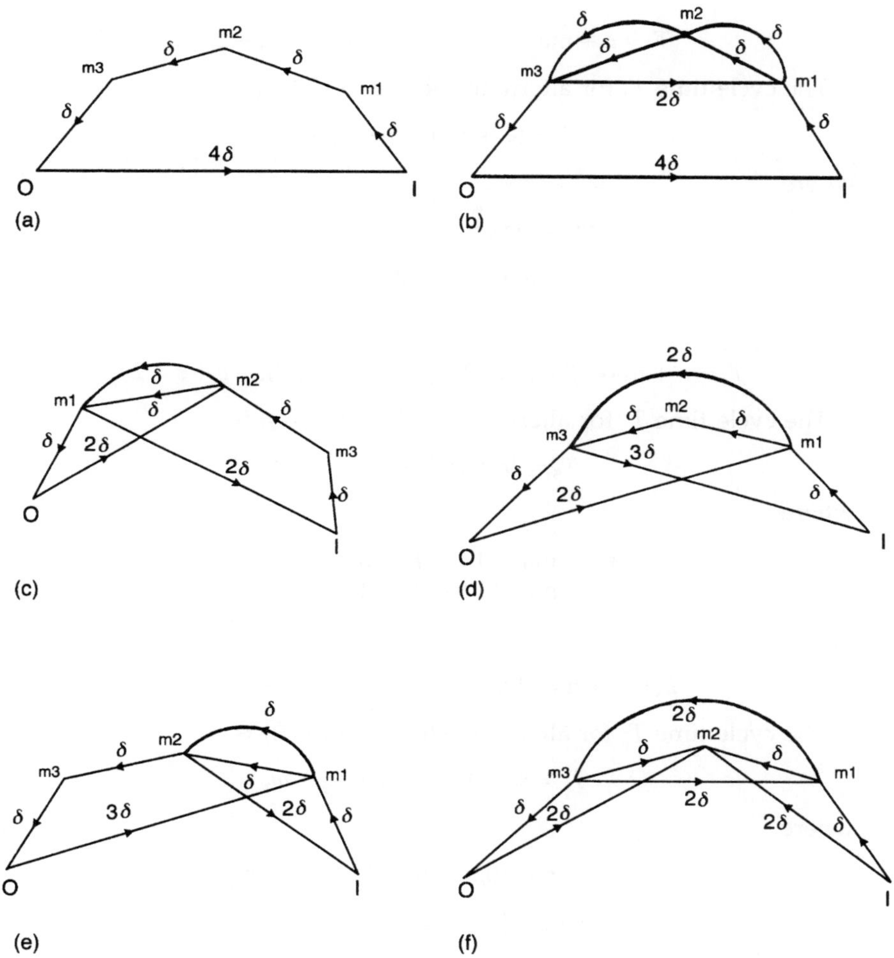

Fig. 8.13 Alternative sequences of robot moves in a three-machine robot cell.

or

$$T_2 = \alpha + \max\{\beta, b, \beta/2 + a, \beta/2 + c, (a+b+c)/2\}$$

The cycle time T_3 for alternative 3 (Fig. 8.13 (c)) is

$$T_3 = 8\varepsilon + 10\delta + w_2 + w_3 + a$$

where

$$w_2 = \max\{0, b - 2\varepsilon - 4\delta - w_3\}$$
$$w_3 = \max\{0, c - 4\varepsilon - 6\delta - a\}$$

or

$$T_3 = a + \max\{\beta - 2\delta + a, c, a + b + \beta/2 - 2\delta\}.$$

The cycle time T_4 for alternative 4 (Fig. 8.13 (d)) is

$$T_4 = 8\varepsilon + 12\delta + b + w_1 + w_3$$

where

$$w_2 = \max\{0, a - 2\varepsilon - 6\delta - w_3\}$$

$$w_3 = \max\{0, c - 2\varepsilon - 6\delta\}$$

or

$$T_4 = \mathrm{a} + \max\{\beta + b,\ \beta/2 + a + b - 2\delta,\ \beta/2 + b + c - 2\delta\}$$

The cycle time T_5 for alternative 5 (Fig. 8.13 (e)) is

$$T_5 = 8\varepsilon + 10\delta + w_1 + w_2 + c$$

where

$$w_1 = \max\{0, a - 4\varepsilon - 6\delta - w_2 - c\}$$
$$w_2 = \max\{0, b - 2\varepsilon - 4\delta\}$$

or

$$T_5 = \mathrm{a} + \max\{a, b + c - 2\delta,\ \beta/2 + b + c - 2\delta\}$$

The cycle time T_6 for alternative 6 (Fig. 8.13 (f)) is

$$T_6 = 8\varepsilon + 12\delta + w_1 + w_2 + w_3$$

where

$$w_1 = \max\{0, a - 4\varepsilon - 8\delta - w_2 - w_3\}$$

$$w_2 = \max\{0, b - 4\varepsilon - 8\delta - w_3\}$$

$$w_3 = \max\{0, c - 4\varepsilon - 8\delta\}$$

or

$$T_6 = \alpha + \max\{\beta, a, b, c\}.$$

From the above alternatives, it is easily seen that alternative 6 dominates alternatives 2 and 4, which are therefore ignored. These results can be represented in an algorithmic form similar to the two-machine case.

Algorithm

Step 0. Calculate $\beta = 4\varepsilon + 8\delta$.
Step 1. If $\beta \geqslant \max\{a, b, c\}$, then T_6 is optimal. Calculate T_6 and stop,

otherwise go to step 2.
Step 2. If $\beta > \max\{a,b,c\}$ and one of the following conditions holds:

(a) $a \geqslant 2\delta$ and $c \geqslant 2\delta$
(b) $a \geqslant 2\delta, c < 2\delta$ and $b + c \geqslant \beta/2 + 2\delta$
(c) $a < 2\delta, c \geqslant 2\delta$ and $a + b \geqslant \beta/2 + 2\delta$
(d) $a < 2\delta, c < 2\delta, a + b \geqslant \beta/2 + 2\delta$ and $b + c \geqslant \beta/2 + 2\delta$

then T_6 is optimal. Calculate T_6 and stop, otherwise go to step 3.
Step 3. If $\beta > \max\{a,b,c\}$ and one of the following conditions holds:

(a) $a \geqslant 2\delta, c < 2\delta$ and $b + c < \beta/2 + 2\delta$
(b) $a < 2\delta, c < 2\delta, a + b \geqslant \beta/2 + 2\delta$ and $b + c < \beta/2 + 2\delta$

then T_5 is optimal. Calculate T_5 and stop, otherwise go to step 4.
Step 4. If $\beta > \max\{a,b,c\}$ and one of the following conditions holds:

(a) $a < 2\delta, c \geqslant 2\delta$ and $a + b < \beta/2 + 2\delta$
(b) $a < 2\delta, c < 2\delta, a + b < \beta/2 + 2\delta$ and $b + c \geqslant \beta/2 + 2\delta$

then T_3 is optimal. Calculate T_3 and stop, otherwise go to step 5.
Step 5. If $\beta > \max\{a,b,c\}$ and $b + c \leqslant 2\delta$ and $a + b \geqslant 2\delta$, then T_1 is optimal. Calculate T_1 and stop.

Example 8.3

Determine the optimal cycle time and corresponding robot sequence for a three-machine robotic cell. The following data are given (adopted from Singh, 1996; reproduced with the permission of John Wiley & Sons, Inc., New York).

processing time for machine m1 = 12.00 min
processing time for m2 = 07.00 min
processing time for m3 = 09.00 min
robot gripper pickup time = 0.19 min
robot gripper release time = 0.19 min
robot move time between two consecutive machines = 0.27 min

Step 0. $\beta = 4\varepsilon + 8\delta = 4(0.19) + 8(0.27) = 2.92$ min
Step 1. $\beta \leqslant \max\{a,b,c\}$, i.e. $2.92 \leqslant \max\{12,7,9\}$. 2.92 is less than 12, thereforeT_6 is optimal. $T_6 = \alpha + \max\{\beta,a,b,c\} = [4\varepsilon + 4\delta] + \max\{2.92,12,7,9\} = [4(0.19) + (0.27)] + 12 = 1.84 + 12 = 13.84$ min
The optimal cycle time is 13.84 min and the optimal robot sequence is given by Fig. 8.13(f).

8.4 SUMMARY

Layout planning is an important aspect of designing cellular manufacturing systems. A conceptual understanding of various issues in layout planning has been provided. Given the complexity of developing a layout process, it is difficult to recommend a single mathematical model that can be used to solve a layout problem. Similarly, traditional approaches to layout planning have several shortcomings. Under these circumstances what is needed is a hybird scheme in which the strength of both approaches can be combined.

PROBLEMS

8.1 Layout considerations in cellular manufacturing systems differ from other manufacturing systems. Discuss the following types of layouts in the design of cellular manufacturing systems: linear single row machine layout; double row machine layout; circular machine layout; cluster machine layout; loop layout.

8.2 Consider five machines in a flexible cellular manufacturing system to be served by an AGV. A linear single row layout is recommended owing to the use of an AGV. Data on the frequency of AGV trips, material handling costs per unit distance, and the clearance between the machines are given in Fig. 8.14 and Table 8.2. Suggest a suitable layout.

8.3 Determine the optimal cycle time as well as the robot sequences in the case of a two-machine robotic cell with the following data:

processing time of m1 = 20.00 min
processing time of m2 = 10.00 min
robot gripper pickup = 0.20 min
robot gripper release time = 0.20 min
robot move time between the two machines = 0.30 min

Now consider another layout arrangement in which two robots are used: one serves one machine and the other robot serves the other machine. Determine under what conditions one robot layout is better than the one with two robots.

8.4 Determine the optimal cycle time and corresponding robot sequence for a three-machine robotic cell. The following data are given:

Table 8.2 Machine dimensions

Machine	*m1*	*m2*	*m3*	*m4*	*m5*
Machine sizes	20 × 10	25 × 15	20 × 30	20 × 10	15 × 25

(a)

	1	2	3	4	5
1	0	30	80	60	40
2	30	0	20	40	25
3	80	20	0	38	51
4	60	40	38	0	55
5	40	25	51	55	0

(b)

	1	2	3	4	5
1	0	3	7	6	4
2	3	0	4	7	2
3	7	4	0	1	9
4	6	7	1	0	2
5	4	2	9	2	0

(c)

	1	2	3	4	5
1	0	1	2	1	2
2	1	0	1	1	2
3	2	1	0	1	2
4	1	1	1	0	1
5	2	2	2	1	0

Fig. 8.14 (a) Frequency of AGV trips; (b) material handling cost; (c) clearance between machines.

processing time for m1 = 10.00 min
processing time for m2 = 07.00 min
processing time for m3 = 04.00 min
robot gripper pickup = 0.20 min
robot gripper release time = 0.20 min
robot move time between two consecutive machines = 0.30 min

Consider several layout arrangements: the first machine is served by the first robot, and the other machines are served by the other robot; the first two machines are served by one robot and the last one is served by another robot; three robots serve one machine each. Determine the conditions under which a given layout is superior to the others.

REFERENCES

Apple, J.M. (1977) *Plant Layout and Material Handling*, Wiley, New York.
Asfahl, C.R. (1992) *Robots and Manufacturing Automation*, Wiley, New York, pp. 272–81.

Askin, R.G. and Standridge, C.R. (1993) *Modeling and Analysis of Manufacturing Systems*, John Wiley & Sons Inc., New York.

Burkard, R. E. (1984) Location with spatial interaction–quadratic assignment problem, in *Discrete Location Theory* (eds. R. L. Francies and P. B. Mischandani), Academic Press, New York.

Das, C. S. and Heragu, S. S. (1995) *Design, Layout and Location of Facilities*, West Educational Publishing, Amesbury, MA.

Francis, R. L., McGinnis, L. F. and White, J. A. (1992) *Facility Layout and Location*, Prentice-Hall Inc., Englewood Cliffs, NJ.

Hales, H.L. (1984) *Computer-Aided Facilities Planning*, Marcel Dekker.

Heragu, S.S. and Kusiak, A. (1988) Machine layout problem in flexible manufacturing systems. *Operations Research*, **36** (2), 258–68.

Heragu, S.S. and Kusiak, A. (1991) Efficient models for facility layout problem. *European Journal of Operational Research*, **53** (1), 1–13.

Hourtzeel, A. and Brown, C. S. (1984) A management overview of group technology, in *Group Technology at Work* (ed. N. L. Hyer), Society of Manufacturing Engineers, Dearborn, MI.

Hyer, N. and Wemmerlöv, U. (1982) MRP/GT: A framework for production planning and control of cellular manufacturing systems. *Decision Science*, **13** (4), 681–701.

Koopmans, T.C. and Beckmann, M. (1957) Assignment problems and location of economic activities. *Econometrica*, **25** (1), 53–76.

Kouvelis, P. and Kim, M.W. (1992) Unidirectional loop network layout problem in automated manufacturing systems. *Operations Research*, **40**(3), 533–50.

Kusiak, A. (1990) *Intelligent Manufacturing Systems*, Prentice-Hall, Englewood Cliffs, NJ.

Kusiak, A. and Heragu, S.S. (1987) The facility layout problem: an invited review. *European Journal of Operation Research*, **29**, 229–51.

Lee, C.R. and Moore, M.J. (1967) CORELAP–computerized relationship layout planning. *Journal of Industrial Engineering*, **18** (3), 195–200.

Leung, J. (1992) A graph theoretic heuristic for designing loop layout manufacturing systems. *European Journal of Operational Research*, **57**(2), 243–52.

Muther, R. (1973) *Systematic Layout Planning*, Van Nostrand Reinhold, New York.

Neghabat, F. (1974) An efficient equipment layout algorithm. *Operations Research*, **22**, 622–8.

Reed, R. (1961) *Plant Layout: Factors, Principles, and Techniques*, Richard D. Irwin, Homewood, IL.

Sethi, S.P., Srikandarajah, C., Blazewicz, J. and Kubiak, W. (1992) Sequencing of robot moves and multiple parts in a robotic cell. *International Journal of Flexible Manufacturing Systems*, **4**, 331–58.

Singh, N. (1996) *Systems Approach to Computer-integrated Design and Manufacturing*, John Wiley & Sons, Inc., New York.

Steudel, H.J. and Desruelle, P. (1992) *Manufacturing in the Nineties: How to Become a Mean, Lean, World-class Competitor*. Van Nostrand Reinhold, New York.

Suer, G.A. and Ortega, M. (1994) Flexibility considerations in designing manufacturing cells: a case study. Univ. Puerto Rico–Mayaguez. Working paper.

Van Camp, D.J., Carter, M.W. and Vannelli, A. (1992) A nonlinear optimization approach for solving facility problem. *European Journal of Operational Research*, **57** (2), 174–89.

Viswanadham, N. and Narahari, Y. (1992) *Performance Modeling of Automated Manufacturing Systems*, Prentice Hall, Englewood Cliffs, NJ.

CHAPTER NINE

Production planning in cellular manufacturing

Formation of cells is an important aspect of cellular manufacturing systems. Once the cells are formed, production planning is the next core activity to realize the benefits of cells. Production planning is concerned with establishing production goals over the planning horizon. The main objective of production planning in any organization is to ensure that the desired products are manufactured at the right time, in the right quantities, meeting quality specifications at minimum cost. A lot of information is required to develop a production plan. This input can then be transformed, using planning tools and techniques, into desirable outputs, that is the production planning process can be conceived as a transformation process in an input–output system framework. Johnson and Montgomery (1974) advocated input–output concepts which are equally valid in a cellular manufacturing environment. These concepts are adopted here with suitable modifications where necessary.

Information required as an input to develop a production plan includes:

- forecasts of future demand;
- alternate process routes for each product/component;
- production standards such as setup information for each machine and variable processing time;
- the capacity of available resources including jigs, fixtures, pallets, material handling equipment and machine tools;
- current inventory levels and the backlog position for each product;
- current work-in-process;
- workforce levels;
- material availability;
- cost standards and selling prices;
- management policies such as overtime, subcontracting and multiple shift operations.

The production planning activity involves transforming these inputs using analytical and logical models. Accordingly, in each period of the planning horizon, the expected output from this production planning activity may take a variety of forms. Typical outputs for each planning period include:

- the number of units of each product to be produced;
- the number of units of each product to be produced by each of the available alternative processes;
- target inventory levels of each product;
- workforce levels;
- overtime, additional shifts and unused capacity;
- quantities of material to be transported within and among cells;
- subcontracting plans;
- purchased material requirements.

The inputs and outputs may differ from period to period, from cell to cell and from one organization to another. One thing is clear, however; production planning involves a myriad of activities. A production planning and control system encompasses in an integrated manner such activities as forecasting end-item demand, translating end-item demand into feasible production plans, establishing detailed planning of material flows and the capacity to support the overall production plans (Singh, 1996). A production plan is a result of interactions among these activities. A brief discussion of a general framework for a manufacturing planning and control system for a batch manufacturing environment, also known as a material requirements planning MRP-based system as suggested by Singh (1996), is provided. A combined GT and MRP framework which exploits the strengths of both systems is also presented.

9.1 A BASIC FRAMEWORK FOR PRODUCTION PLANNING AND CONTROL

A production planning and control system (PPCS) is required for the reasons outlined above and helps execute the plans by such actions as detailed cell scheduling and purchasing. A number of benefits arise from the use of an integrated PPCS:

- reduced inventories;
- reduced capacity;
- reduced labor and overtime costs;
- shorter manufacturing lead times;
- faster responsiveness to internal and external changes such as machine and other equipment failures, product mix and demand changes etc.

This section adopts the basic framework for developing a PPCS from Singh (1996). The major elements of an integrated PPCS are:

- demand management
- aggregate production planning
- master production schedule
- rough-cut capacity planning
- material requirements planning
- capacity planning
- order release
- shopfloor scheduling and control

The flow of information among various elements of a PPCS is given in Fig. 9.1. The demand for products is the driving force behind any manufacturing activity, so the demand management module is one of the most important elements of a PPCS. The primary function of the demand management module is demand forecasting. It not only

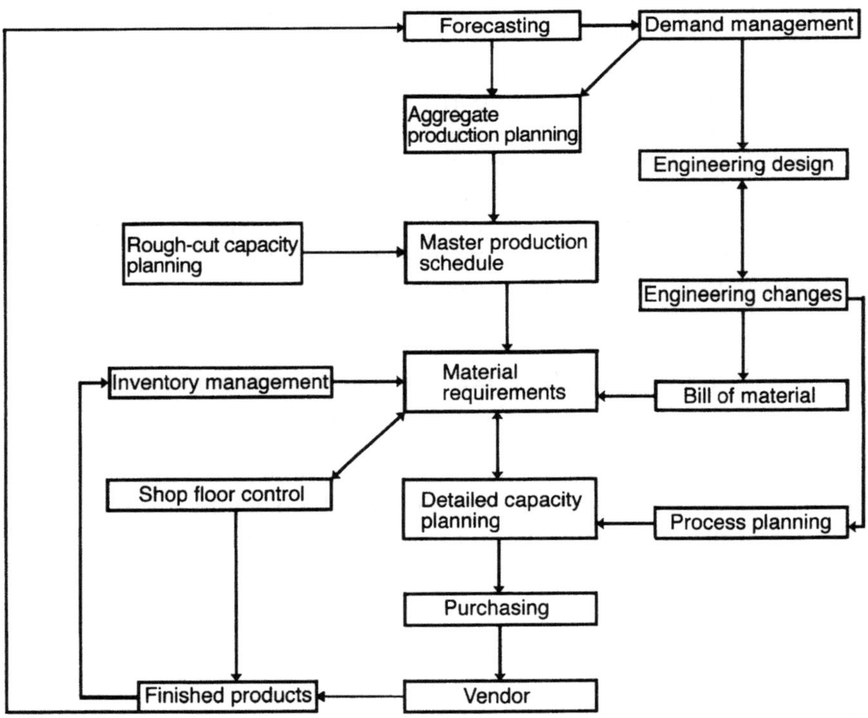

Fig. 9.1 Basic framework for a planning and control system. (Source: Singh N., *Systems Approach to Computer-integrated Design and Manufacturing*, © 1996. Reprinted by permission of John Wiley & Sons, Inc., New York.)

provides an estimate of the demand for each type of product, it also provides a link between the PPCS and the marketplace. It helps establish a channel of communication between manufacturer and customers.

The physical resources of firms are normally fixed during the planning horizon. If there are a number of product types with time-varying demands, which is normally the case in discrete product manufacturing environments, then the production should be planned in an aggregate manner to utilize the resources effectively. Demand forecasting is an important input to aggregate production planning. The objective of aggregate production planning is to rationalize the differences between the forecast demand for products and capacity over the planning horizon. In aggregate production planning, the demand and production requirements are represented in common aggregate units such as plant hours or direct labor hours. The aggregate production plan must be disaggregated to determine the quantity of each product to be produced in each period during the planning horizon. Such a disaggregate plan for each product is known as a master production schedule (MPS).

The feasibility of an MPS has to be assured, based on rough-cut capacity planning. However, the final assembly of each end-item (product) consists of a number of sub-assemblies and several components. Further, in a real-life manufacturing environment, inventories exist for some of the components and sub-assemblies. Under these circumstances the MPS cannot be used for developing detailed production plans for end-items.

The MRP system is used to determine the detailed production plans. The MPS together with information on on-hand stock, purchased and manufacturing order status, order quantity, lead time and safety stock and product structure are inputs to the MRP system. Output from MRP determines how many of each item from the bill of materials must be manufactured in each period.

It may so happen that the production plan suggested by the MRP system may exceed available capacity for some of the components. Such infeasibilities are determined by what is known as the detailed capacity analysis. There are a number of ways to resolve capacity limitation problems. For example, possible alternatives are multiple shifts, overtime, subcontracting, varying the production rate by hiring and lay-offs, and building inventories. If these solutions do not work, there is no alternative but to modify the MPS. Every modification in the MPS results in changes in MRP calculations and, consequently, the MPS and MRP are, most of the time, iterative processes.

Once the feasibility of the detailed production plan is assured, the next step in the process is known as order release. This refers to the process of issuing directives to work, which means releasing production orders for the parts to be manufactured, and purchase orders for the parts to be purchased. Once the orders are released, the next most complex step is production control.

Many random and complex events take place once production begins. For example, tools and machines exhibit random failure phenomena causing variations in the production rates; the work centers may starve due to the non-arrival of parts ordered in time due to the inherent uncertainties in the purchase process. The objective of the production control function is to accommodate these changes by scheduling work orders on the work centers, sequencing jobs in a work order at a work center, and monitoring purchase orders from vendors. The activities of scheduling and sequencing are known as shopfloor control. Once all the items are manufactured they are used in sub-assemblies and then assemblies. The final assemblies can be shipped to the customers according to the shipping schedule.

The framework outlined forms a foundation for an integrated PPCS. The detailed system may, however, differ from one organization to another. The following sections discuss some of the elements of a PPCS.

Aggregate production planning

In a high variety, discrete product manufacturing environment, the demand for products fluctuates considerably. However, the resources of a firm, such as the capacity of its machines and workforce, normally remain fixed during the planning horizon of interest, which varies from 6 to 18 months; 12 months is a suitable period for most PPCSs. Under such circumstances (a large variety of products, time-varying demand, a long planning horizon and fixed available resources) the best approach to obtain feasible solutions is to aggregate the information being processed, for example, by grouping similar items into product families, grouping machines processing these products into machine cells and grouping workers with different skills into labor centers. For aggregation purposes the product demand should be expressed in a common unit of measurement such as production hours, plant hours etc.

Since production planning is primarily concerned with determining optimal production, inventory and workforce levels to meet demand fluctuations, a number of strategies are available to management to absorb the demand fluctuations:

- maintain a uniform production rate and absorb demand fluctuations by accumulating inventories;
- maintain a uniform workforce but vary the production rate by permitting planned overtime, idle time and subcontracting;
- change the production rate by changing the size of the workforce through planned hiring and lay-offs;

- explore the possibility of planned backlogs if customers are willing to accept delays in filling their orders.

A suitable combination of these strategies should be explored to develop an optimal aggregate production plan. First, the implications of various alternative aggregate production plans will be illustrated with a simple example, and then a simple linear programming formulation for determining an optimal aggregate production plan will be presented.

Example 9.1

Data on the expected aggregated sales of three products A, B and C over six four-week planning horizons are given in Table 9.1(a), and the aggregate demand forecast in cell-hours is given in Table 9.1(b). The company has developed machining-cell hours as a common unit for aggregation purposes. In this case product A and B require two cell-hours per unit whereas product C requires only one cell-hour per unit. The company has a regular production capacity of 300 units per period which can be varied up to 350 units per period. Overtime is permitted up to a maximum of 60 units per period. Requirements exceeding overtime capacity can be satisfied by subcontracting. Two alternate production policies are developed as follows:

plan I: produce at the constant rate of 350 units per period for the entire planning horizon (Table 9.1 (c));
plan II: produce at the rate of 400 units per period for the first four periods and then at the rate of 250 units per period for the subsequent periods (Table 9.1 (d)).

The following is an analysis of two aggregate production plans suggested by the production department of Windsor Steel Manufacturing Company.

It is assumed that there is no initial inventory and all shortages are back-ordered. Further, it is assumed that the regular time capacity can be varied up to 300 units per period. Under these conditions, plan I results in back-orders in three periods with very little inventory. However, plan II eliminates the backlogs and results in more overtime and subcontracting; inventory build-up is higher. The change in production capacity is measured from the regular production capacity. For example, in the case of plan I the production rate is 350 whereas the regular production capacity is 300. Therefore, the change in capacity is +50 units. Similarly, in the case of plan II, the change in capacity in period 5 from the regular production capacity of 300 to 250 is only −50 units and not −150 units which appears when the capacity is changing from 400 to 250 between periods 4 and 5.

Table 9.1 Forecast for six four-week periods: (a) Demand forecast in units for products A, B and C; (b) Expected aggregate demand; (c) Plan I, uniform regular production rate policy; (d) Plan II, varying regular production rate policy. (Source: Singh, N. *Systems Approach to Computer Integrated Design and Manufacturing*, 1996. Reproduced with permission from John Wiley & Sons, Inc., New York.)

Period	*Product A*		*Product B*		*Product C*	
	Units	*Equivalent cell-hours*	*Units*	*Equivalent cell hours*	*Units*	*Equivalent cell hours*
1	60	120	40	80	100	100
2	70	140	50	100	160	160
3	50	100	70	140	210	210
4	55	110	65	130	170	170
5	45	90	55	110	100	100
6	40	80	40	80	80	80

Period	*Expected aggregate (equivalent cell-hous)*	*Cumulative aggregate (cell-hours)*
1	300	300
2	400	700
3	450	1150
4	410	1560
5	300	1860
6	240	2100

Period	*Production Rate*	*Inventory*	*Back-orders*	*Change in Capacity*	*Overtime*	*Sub-contract*
1	350	50	0	+50	50	0
2	350	0	0	0	50	0
3	350	0	100	0	50	0
4	350	0	160	0	50	0
5	350	0	110	0	50	0
6	350	0	0	0	50	0

Period	*Production Rate*	*Inventory*	*Back-orders*	*Change in Capacity*	*Overtime*	*Sub-contract*
1	400	100	0	+100	60	40
2	400	100	0	0	60	40
3	400	50	0	0	60	40
4	400	40	0	0	60	40
5	250	0	10	−50	0	0
6	250	0	0	0	0	0

Two alternatives were analysed. There could, however, be a large number of alternative aggregate production plans. The question is which one is the best considering all the relevant costs and system constraints. The following section presents a mathematical programming model to obtain an optimal aggregate production plan.

Mathematical programming model

Several mathematical models have been developed which seek an optimal combination of various strategies outlined earlier (for details on a number of models, see Johnson and Montgomery, 1974; Hax and Candea 1984). The following is a simple linear programming model for this purpose.

Minimize

$$Z = \sum_{t=1}^{T} \{C_x X_t + C_w W_t + C_o O_t + C_u U_t + C_i I_t^+ + C_s I_t^- + C_h H_t + C_f F_t\} \tag{9.1}$$

subject to:

$$X_t + I_{t-1}^+ - I_{t-1}^- - I_t^+ + I_t^- = d_t, \quad \forall\, t, t = 1, 2, \ldots, T \tag{9.2}$$

$$W_t = W_{t-1} + H_t - F_t \tag{9.3}$$

$$O_t - U_t = kX_t - W_t \tag{9.4}$$

$$X_t, W_t, O_t, U_t, I_t^+ I_t^-, H_t, F_t \geqslant 0 \tag{9.5}$$

Notation

C_f cost of lay-offs per hour
C_h cost of hiring regular-time labor per hour
C_i cost of unit inventory holding per hour
C_o overtime labor cost per hour
C_s cost of a shortage per hour
C_u opportunity cost of not using the equipment
C_w cost of regular-time labor per hour
C_x variable cost of production excluding labor per hour
d_t aggregate demand in period t
F_t lay-offs in hours in period t
H_t new labor force added in hours in period t
I_t^+ inventory in stock at the end of period t
I_t^- shortages at the end of period t
k labor hours required to produce a unit
O_t overtime scheduled in period t

T number of periods in the planning horizon
U_t undertime allowed in period t
W_t regular-time workforce level in period t
X_t production units scheduled in period t
Z total system cost

The objective function is given by equation 9.1 which includes the costs of production, regular-time workforce, overtime, undertime, inventory, shortages, hiring and lay-offs for all the periods in the planning horizon. Equation 9.2 provides the inventory balance considering demand and production in all periods. Labor balance is given by equation 9.3 and the relationship between overtime, undertime, workforce and production is given by equation 9.4. Non-negativity constraints are given by relation 9.5.

Example 9.2
Develop an optimal aggregate production plan in terms of cell-hours for the forecast demand data given in example 9.1. Other data are:

$C_x = \$100$ per hour, $C_o = \$20$ per hour, $C_w = \$14$ per hour, $C_h = \$14$ per hour, $C_f = \$30$ per hour, $C_i = \$3$ per unit per hour, $C_s = \$400$ per unit per hour, $C_u = \$50$ per hour. The initial regular workforce level $W_o = 240$ labor-hours per hour and $k = 1$.

The objective function and the constraints are given in the appendix at the end of this chapter. On solving the linear programming model, the aggregate production plan obtained is as given in Table 9.2 (a). It can be seen that the demand fluctuations are absorbed by using the overtime permitted. Accordingly, no hiring, firing, inventory or shortages are incurred. However, if the overtime is restricted, then the scenario will change. For example, if the overtime is restricted to 50 units in periods 1 to 4, the output of the aggregate production planning model is as given in Table 9.2 (b). It can be seen that the optimal plan is now different and requires the use of overtime, hiring, firing, inventory and back-orders to absorb the demand fluctuations.

Master production schedule (MPS)

When a large number of products are manufactured in a company, the primary use of an aggregate production plan is to level the production schedule such that the production costs are minimized. However, the output of an aggregate production plan does not indicate individual products, so the aggregated plan must be disaggregated into individual products. A number of disaggregation methodologies have been developed (Bitran and Hax, 1977; Bitran and Hax, 1981); disaggregation approaches were compared by Bitran, Hass and Hax (1981). The treatment of these approaches is complex and is beyond the scope of

Table 9.2 (a) Output of aggregate production planing model; (b) Overcome restricted output of aggregate production planning model; (Source: Singh, N. *Systems Approach to Computer Integrated Design and Manufacturing,* 1996. Reproduced with permission from John Wiley & Sons, Inc., New York.)

Period	d_t	X_t	W_t	O_t	H_t	F_t	U_t	I_t^+	I_t^-
1	300	300	240	60	00	00	00	00	00
2	400	400	240	160	00	00	00	00	00
3	450	450	240	210	00	00	00	00	00
4	410	410	240	170	00	00	00	00	00
5	300	300	240	60	00	00	00	00	00
6	240	240	240	00	00	00	00	00	00

(a)

Period	d_t	X_t	W_t	O_t	H_t	F_t	U_t	I_t^+	I_t^-
1	300	290	240	50	00	00	00	00	00
2	400	430	380	50	140	00	00	20	00
3	450	430	380	50	00	00	00	00	00
4	410	410	380	30	00	00	00	00	00
5	300	300	300	00	00	00	00	00	00
6	240	240	240	00	00	00	00	00	00

(b)

this book. The result of such a disaggregation is what is known as a master production schedule. Does the MPS present an executable manufacturing plan? Not really, because the capacities and the inventories have not been considered at this stage. Therefore, further analysis for the material and capacity requirements is necessary to develop an executable manufacturing plan. The following sections discuss the basic concepts of rough-cut capacity planning, MRP and capacity planning, which can be used to develop executable production plans.

Rough-cut capacity planning

The primary objective of rough-cut capacity planning is to ensure that the MPS is feasible. For each product family the average work needed on key work centers per unit can be calculated by the bill of materials and routings for each item. The resource profile defined as the amount of work in some meaningful unit of measure by resource per unit of output is developed. For example, consider two families of steel cylinders and the resource profile developed in standard hours of resources per 200 units of end-product family given in Table 9.3. The

Table 9.3 Resource profile for two product families. (Source: Singh, N. *Systems Approach to Computer Integrated Design and Manufacturing,* 1996. Reproduced with permission from John Wiley & Sons, Inc., New York.)

Work center	*Standard hours per 200 units*		*Total resources required for all families*
	Product family I	*Product family II*	
1100	14	7	21
2100	7	20	27
3100	6	14	20
4500	25	9	34
6500	9	16	25

available resources are compared with the resource requirements profile obtained for all work centers considering all product families. If the available resources are less than those required, then the problem can be resolved by management decisions relative to overtime, subcontracting, hiring, lay-offs, and so on.

Material requirements planning (MRP)

The MRP system is essentially an information system consisting of logical procedures to manage inventories of component assemblies, sub-assemblies, parts and raw materials in a manufacturing environment. The primary objective of an MRP system is to determine time-phased requirements of each item in the bill of materials, that is how much of each item must be manufactured or purchased and when. The key concepts used in determining the material requirements are:

- product structure and bill of materials;
- independent versus dependent demand;
- parts explosion;
- gross requirements;
- scheduled receipts;
- on-hand inventories;
- net requirements;
- planned order releases;
- lead time.

A brief description of these concepts is given below.

Product structure and bill of materials

Product is the single most important identity in any organization. The product is what a company sells to its customers. The survival of a company consequently depends on the profit on the sales of the product. A product may be made from one or more assemblies, sub-assemblies and components. The components are made from some form of raw material. However, the types of raw material, components, sub-assemblies and final assemblies vary from product to product. For example, the types and quantities of raw materials, components and sub-assemblies for a household refrigerator are different from those of a color television. To manufacture the products it is therefore important that the product structure is properly understood and the correct information is available on the components, sub-assemblies and assemblies.

A bill of materials is an engineering document that specifies the components and sub-assemblies required to make each end-item (product). Consider a hypothetical product called end-item E1 (at level 0) which is made up of two sub-assemblies S1 and S2 at level 1, as shown in Fig. 9.2. Each sub-assembly S1 and S2 at level 2 consists of two and three components at level 2 respectively. A complete product structure for product E1 is shown in Fig. 9.2. The end-item E1 is called a parent item to sub-assemblies S1 and S2, which are called component items. Similarly, sub-assembly S1 is a parent item to components C1 and C2, and S2 is a parent item to C3, C4 and C5. At level 3, the raw material is input to the components at level 2.

Independent versus dependent demand

The demand for end-items originates from customer orders and forecasts. Such a demand for end-items and spare parts is called independent demand. However, the demand by a parent item for its

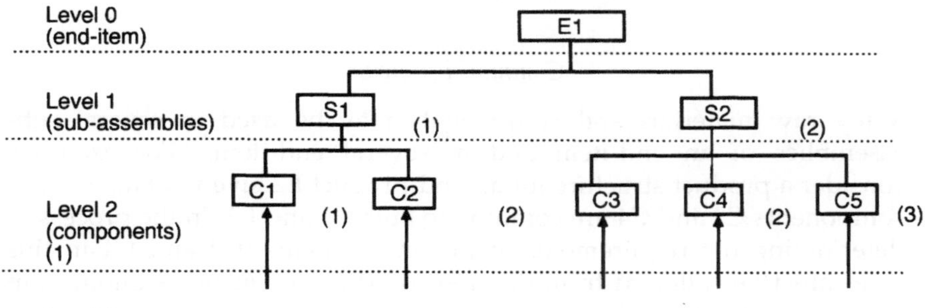

Fig. 9.2 Product structure for hypothetical product and bill of materials. (Source: Singh N., *Systems Approach to Computer-integrated Design and Manufacturing*, © 1996. Reprinted by permission of John Wiley & Sons, Inc., New York.)

components is called dependent demand. For example, if the end-item demand is X units and one unit of end-item requires Y units of sub-assembly, then the demand of that sub-assembly is XY units.

Parts explosion

The process of determining gross requirements for component items, that is, requirements for the sub-assemblies, components and raw materials for a given number of end-item units, is known as parts explosion; the parts explosion essentially represents the explosion of parents into their components.

Gross requirements of component items

To compute gross requirements of component items, it is necessary to know the amounts required of each component item to obtain one parent item. This information is available from the product structure and the bill of materials. For example, if the demand for the end-product E1 from a market survey in period 7 is 50 units (for the product structure and the bill of materials/see Fig. 9.2), then the dependent demand (gross requirements) for the sub-assemblies and components can be determined as follows:

demand of S1 = 1 × demand of E1 = 50 units
demand of S2 = 2 × demand of E1 = 100 units
demand of C1 = 1 × demand of S1 = 50 units
demand of C2 = 2 × demand of S1 = 100 units
demand of C3 = 2 × demand of S2 = 200 units
demand of C4 = 3 × demand of S2 = 300 units
demand of C5 = 1 × demand of S2 = 100 units

Common-use items

Many raw materials and components may be used in several sub-assemblies of an end-item and in several end-items. For example, consider a product structure for an end-product E2 given in Fig. 9.3. The components C2 and C4 are common to both E1 and E2. In the process of determining net requirements, common-use items (C2 and C4 in this case) must be collected from different products to ensure economies in the manufacturing and purchasing of these items.

On-hand inventory, scheduled receipts and net requirements

In some cases when there is ongoing production activity, there is some initial inventory for some of the component items available from the

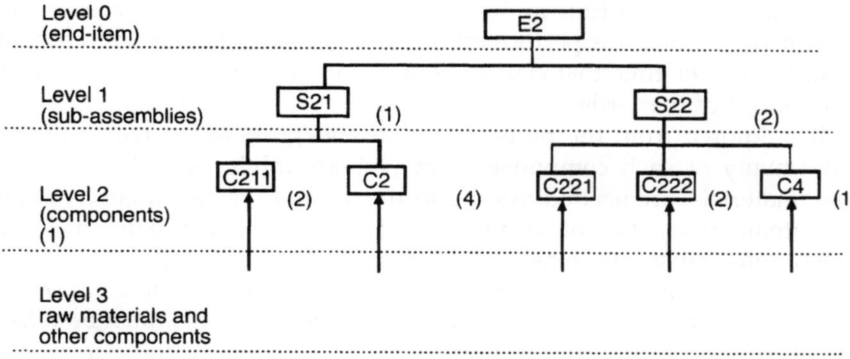

Fig. 9.3 Product structure for end-item E2. (Source: Singh N., *Systems Approach to Computer-integrated Design and Manufacturing*, © 1996. Reprinted by permission of John Wiley & Sons, Inc., New York.)

previous production runs. Also, to maintain continuous production from one planning horizon to another, some inventory is planned to be available at the end of the planning horizon. This inventory is referred to as on-hand inventory for the current planning periods. Further, it takes some time for the orders to arrive. Therefore, the orders placed now are delivered in some future periods. These are known as scheduled receipts. The net requirements in a period are thus obtained by subtracting the on-hand inventory and on-order inventory from the gross requirements.

Planned order releases

Planned order releases refer to the process of releasing one lot of every component item for production or purchase. The question is how to determine the economic lot-sizes of component items. Since in a MRP system shortages are not permitted, the lot-sizes are determined by trading-off the inventory holding cost and setup costs. Although the manufacturing system is a multistage production system, the demand at each stage (level) is deterministic and time-varying. The lot-sizes in a MRP system are determined for component items for each stage sequentially, starting with level 1, then level 2 and so on. A number of lot-sizing techniques have been developed; only Wagner and Whitin (1958) is optimal. A comparative analysis of some heuristic algorithms was given by Naidu and Singh (1987).

Lead time and lead time offsetting

The lead time is how long it takes to produce or purchase a part. In the case of manufacturing, the lead time depends on the setup time, the variable production time and the lot-size, the sequence of machines on which operations are performed, queuing delays etc. The purchasing lead time is the time that elapses between an order with the vendor and the receipt of that order.

It is known from the parts explosion and gross-to-net requirements how many of each component item (sub-assemblies, components and raw materials) are needed to support the desired finished quantity of an end-item. Information on lead time, that is, on the sequence in which the operations must be done and the time it takes to perform these operations on a given lot-size, is required to schedule the component items. The manufacture or purchase of component items must be offset by at least their lead times to ensure the availability of these items for assembly into the parent items at the desired time.

Example 9.3
Consider the product structure of end-item E1 given in Fig. 9.2. The end-item demands from the MPS for weeks 3 to 10 are: 20, 30, 10, 40, 50, 30, 30 and 40 units, respectively. The manufacturing/assembly lead times E1, S2 and C4, the ordering lead time for M4, the on-hand inventory and scheduled receipts are given in Table 9.4. Carry out the MRP procedure for raw material M4 required to manufacture component C4.

The solution is presented in Table 9.4. There is an on-hand inventory of 50 units for end-item E1. The net requirements are obtained by first satisfying the demand from the on-hand inventory. The net requirements for E1 are then offset by three periods to obtain the planned order releases of 10, 40, 50, 30, 30 and 40 in periods 2 to 7, respectively. Two sub-assemblies of S2 are needed to support E1. Accordingly, the gross requirements of S2 are obtained by doubling the planned order releases of E1. The calculations for the net requirements of S2 are offset by two weeks to obtain the planned order releases. A similar process applies to component items C4 and M4 as shown in Table 9.4.

Capacity planning

The planned order releases of all items to be produced during a period are set without considering the available capacity of the work centers. This may lead to an infeasible production plan when the available capacity is less than that required by the MRP plan. Capacity planning is concerned with ensuring the feasibility of the production plans by determining resources such as labor and equipment to develop an executable manufacturing plan. This can be achieved by considering the following alternatives such as overtime, subcontracting, hiring and firing, building inventory and increasing capacity by adding more equipment. If none of these alternatives is sufficient the MPS should be

Table 9.4 MRP planned order releases of component items. (Source: Singh, N. *Systems Approach to Computer Integrated Design and Manufacturing*, 1996. Reproduced with permission from John Wiley & Sons, Inc., New York.)

End-item E1: Leadtime: 3 weeks										
Periods (in weeks)	1	2	3	4	5	6	7	8	9	10
On-hand inventory: 50										
Gross requirements			20	30	10	40	50	30	30	40
Scheduled receipts										
Net requirements					10	40	50	30	30	40
Planned order release		10	40	50	30	30	40			
Component item S-2: Lead time: 2 weeks										
Periods (in weeks)	1	2	3	4	5	6	7	8	9	10
On-hand inventory: 100		20	80	100	60	60	80			
Gross requirements										
Scheduled receipts	100	30								
Net requirements					30	60	80			
Planned order release			30	60	80					
Component item C4: Lead time: 2 weeks										
Periods (in weeks)	1	2	3	4	5	6	7	8	9	10
On-hand inventory: 50			90	180	240					
Gross requirements										
Scheduled receipts	50	20								
Net requirements				150	240					
Planned order release			150	240						
Component item M4: Lead time: 1 week										
Periods (in weeks)	1	2	3	4	5	6	7	8	9	10
On-hand inventory: 150			300	480						
Gross requirements										
Scheduled receipts	50			300						
Net requirements			100	180						
Planned order release		100	180							

modified. This process should continue until the feasibility of production plans is assured.

Shop floor control

After ensuring the feasibility of the detailed production plans the next step should be releasing production orders to be manufactured and the purchase orders to be purchased. In real production many random events occur such as machine failure and uncertainties in supplies; thus, to accommodate these uncertainties production control offers an efficient way to schedule job orders on the work centers and to sequence the jobs on a work center.

9.2 PRODUCTION PLANNING AND CONTROL IN CELLULAR MANUFACTURING SYSTEMS

This section discusses some issues related to production planning and control in cellular manufacturing systems. The basic framework proposed in the previous section suggests an hierarchical decision process for manufacturing planning and control. This framework is also applicable to cellular manufacturing systems with suitable modifications. These modifications take advantage of similarities in setups and operations by integrating GT concepts with MRP. An hierarchical approach for cell planning and control by integrating the concepts of GT and MRP is given here. Together with examples, it is shown how the concepts of GT and MRP can be used together to provide an efficient tool for production planning and control in cellular manufacturing.

It is known that GT is a useful approach to small-lot multi-product production. It is also known that MRP is an effective PPCS for a batch-type production system. In a MRP-based system, optimal lot-sizes are determined for various parts required for products. However, similarities among the parts requiring similar operations are not exploited. The grouping of parts for loading and scheduling based on similiar setups and operations will reduce setup time. On the other hand, the time-phased requirement planning aspect is not considered in GT. This means that all the parts in a group are assumed to be available at the beginning of the period. Obviously, the integration of GT and MRP will lead to better PPCSs. We provide an integrated GT and MRP framework for production planning and control in cellular manufacturing systems (Ham, Hitomi and Yoshida, 1985; Singh, 1996).

Integrated GT and MRP framework

The objective of an integrated GT and MRP framework is to exploit the similarities of setups and operations from GT and time-phased requirements from MRP. This can be accomplished through a series of simple steps (Ham, Hitomi and Yoshida, 1985):

Step 1: Gather the data normally required for both the GT and MRP concepts (that is, parts and their description, machine capabilities, a breakdown of each final product into its individual components, a forecast of final product demand, etc.).

Step 2. Use GT procedures discussed in previous chapters to determine part families. Designate each family as $GI\,(I = 1, 2, \ldots, N)$.

Step 3: Use MRP to assign each component part to a specific time period.
Step 4: Arrange the component part/time period assignments of step 3 according to the part family groups of step 2.
Step 5: Use a suitable group scheduling algorithm to determine the optimal schedule for all those parts within a given group for each time period.

The following simple example illustrates the integrated framework.

Example 9.4
Johnson and Johnson (JJ) produces all the parts in a flexible manufacturing cell required to assemble five products designated P1–P5. These products are assembled using parts A1–A9. The product structure is given in Table 9.5. Using GT, these nine parts can be divided into three part families, designated G1–G3. The number of units required for each product for the month of March have been determined to be: P1 = 50, P2 = 100, P3 = 150, P4 = 100 and P5 = 100. This demand is further exploded to parts level and the information is summarized in Table 9.6. However, if a group scheduling algorithm alone is used on the data given in Table 9.6, such a schedule could well violate specific due-date constraints. For example, one might schedule 50 units of product P1 for production in week 2, whereas 25 of these are actually needed in week 1. Using MRP, the precise number of each part on a short-term (e.g. weekly) basis may be determined. Table 9.7 illustrates

Table 9.5 Product structure. (Source: Singh, N. *Systems Approach to Computer Integrated Design and Manufacturing,* 1996. Reproduced with permission from John Wiley & Sons, Inc., New York.)

Product name	*Part name*	*Number of units required*
P1	A1	1
	A2	1
	A3	2
P2	A2	1
	A4	1
	A6	1
P3	A1	1
	A2	1
	A5	1
P4	A6	1
	A7	1
	A8	1
P5	A7	1
	A8	1
	A9	1

such an MRP output for this example, giving the number of units of each product needed in each week of the month under consideration. However, the optimal schedule within each week is not known. Thus, to take full advantage of the integrated GT/MRP system, Tables 9.5–9.7 are combined in the integrated form given in Table 9.8. Next, by applying an appropriate scheduling algorithm to these sets of parts within a common group and week, an optimal schedule may be obtained for each week of the entire month that takes advantage of GT-induced cellular manufacturing as well as the MRP derived due-date considerations. This is illustrated in the following example.

Example 9.5

Three groups of parts discussed in Example 9.4 are to be manufactured on a machining center in a flexible manufacturing cell which has multiple spindles and a tool magazine with 150 slots for tools. Group setup time and unit processing time for all the parts are given in Table 9.9. The machining center is available for 1020 units of time per week. Using the data given in this and Example 9.4: determine if the available capacity is sufficient for all the weeks; determine the scheduling sequence for groups and parts within each group.

To assess capacity, using the data of Tables 9.8 and 9.9 the capacity required for processing parts in group 1 in the first week can be calculated as follows:

Group setup time of G1 + week 1 demand × unit processing time of A1 + week 1 demand × unit processing time of A3 + week 1 demand × unit processing time of A5 $= 15 + 50 \times 2 + 50 \times 3 + 25 \times 4 = 365$

Similarly, the capacity requirements for all other groups in all the weeks can be calculated. The results are summarized in Table 9.10. It can be observed that the given capacity of 1020 units per week is sufficient for

Table 9.6 Monthly parts requirement in each group (Reproduced from Singh, 1996. Printed with permission of John Wiley & Sons, Inc., New York.)

Group	*Part name*	*Monthly requirement*
G1	A1	200
	A3	100
	A5	150
G2	A2	300
	A4	100
G3	A6	200
	A7	200
	A8	200
	A9	100

Table 9.7 Planned order releases for the products (Reproduced from Singh, 1996. Printed with permission of John Wiley & Sons, Inc., New York.)

Part name	*Week 1*	*Week 2*	*Week 3*	*Week 4*
P1	25	00	25	00
P2	25	25	25	25
P3	25	50	25	50
P4	50	00	00	50
P5	00	50	50	00

Table 9.8 Combined GT/MRP data (Reproduced from Singh, 1996. Printed with permission of John Wiley & Sons, Inc., New York.)

		Planned order release for parts			
Group	*Part name*	*Week 1 demand*	*Week 2 demand*	*Week 3 demand*	*Week 4 demand*
G1	A1	50	50	50	50
	A3	50	00	50	00
	A5	25	50	25	50
G2	A2	75	75	75	75
	A4	25	25	25	25
G3	A6	75	25	25	75
	A7	50	50	50	50
	A8	50	50	50	50
	A9	00	50	50	00

Table 9.9 Group setup and unit processing time for all the parts (Reproduced from Singh, 1996. Printed with permission of John Wiley & Sons, Inc., New York.)

Group name	*Group setup time*	*Parts name*	*Unit processing time*
G1	15	A1	2
		A3	3
		A5	4
G2	10	A2	3
		A4	4
G3	20	A6	2
		A7	3
		A8	2
		A9	1

only week 2. For the remaining weeks decisions have to be made about overtime or subcontracting or some other policies to meet capacity requirements. Cost information about overtime and subcontracting may be helpful in making these decisions.

Table 9.10 Capacity requirements for part groups (Reproduced from Singh, 1996. Printed with permission of John Wiley & Sons, Inc., New York.)

Group name	*Week 1*	*Week 2*	*Week 3*	*Week 4*
G1	365	315	365	315
G2	335	335	335	335
G3	420	370	370	420
Total capacity	1120	1020	1070	1060

Regarding the scheduling sequence, suppose the objective is to minimize the mean completion time of all the parts in the cell. One simple and efficient way to schedule is the shortest processing time (SPT) rule. To sequence groups consider the total processing time required by all the jobs in the group and the group setup time. These times are given in Table 9.9. Accordingly, using the SPT rule, the following sequences may be scheduled:

Week 1: G2, G1 and G3
Week 2: G1, G2 and G3
Week 3: G2, G1 and G3
Week 4: G1, G2 and G3.

Similarly, parts within a group can be sequenced using SPT. For the first week, the parts in G1 will be sequenced in the order A1, A5 and A3, the sequence of the parts in group G2 will be A4 and A2, and the sequence of parts in group G3 will be A7, A8 and A6. Sequences for the weeks 2–4 can be decided similarly.

Period batch control approach

Burbidge (1975) suggested period batch control (PBC) as a proper tool for production planning and control in cells. The PBC approach is quite similar to the integrated GT/MRP approach. The major difference is that the PBC is a single-cycle system that starts by dividing the planning period into cycles of equal length. Hyer and Wemmerlov (1982) presented the following hierarchical decision process based on the PBC concept of a single cycle:

Level 1: The planning horizon is divided into cycles of equal length, say *n* weeks. Based on a sales forcast, the MPS is generated for end-items in each cycle.

Level 2: The MPS in a specific cycle is exploded into its parts requirements by using a list order form analogous to a bill of materials. Lot-for-lot sizing procedures are used for component parts.

Level 3: All the parts scheduled for production in a given cycle are categorized by family. The families formed by similarities in processing requirements are assigned to cells with the required capabilities. Planned loading sequences created to take advantage of similar tooling requirements are used to sequence the jobs into the cells.

PBC is single-cycle because all parts are ordered with a frequency determined by the same time interval, the cycle; it is single-phase in that all parts have the same planned start date (the beginning of the cycle) and the same due date (the end of the cycle). Figure 9.4 illustrates the PBC cyclical approach to production planning. The lot-sizing approach for these component parts is lot-for-lot where the lot-size for any part is the number of parts required for the cycle. All these component parts will be processed in the 'make' cycle preceding the assembly cycle. All these quantities are then ordered to a 'standard schedule', which allows time for raw material production or delivery, component processing and assembly. This standard schedule is repeated every cycle.

Planning the loading sequence

Cell formation provides the best division of the plant into groups of machines and families of parts to be processed in each group (cell). PBC, in turn, finds for each group how many of each part should be produced during each cycle to meet the demand and utilize the available capacity. Another important issue is the dispatching strategy which determines

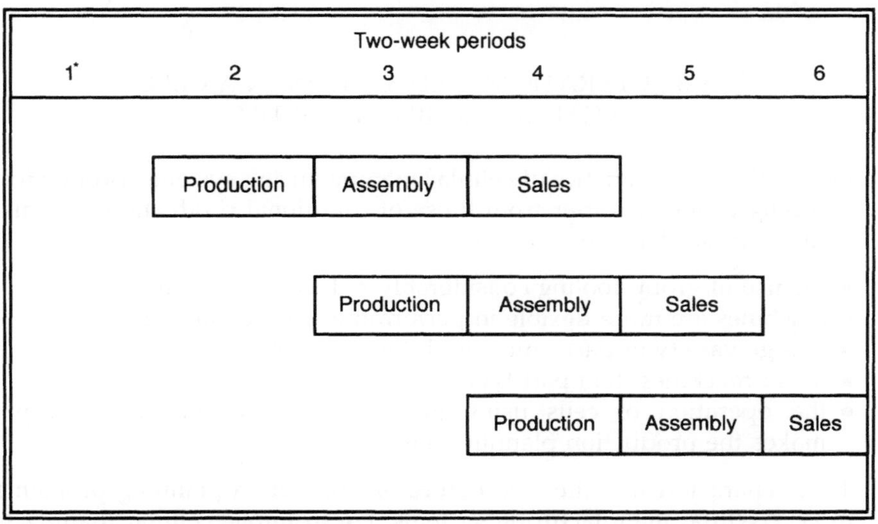

Fig. 9.4 Single-cycle approach of PBC.

the sequence for loading the parts on the machines in a group each cycle. This issue is related to production control. Generic issues of production control are discussed in the next chapter.

Advantages of using PBC

A number of advantages of using PBC have been documented (Burbidge, 1975).

1. The single-cycle ordering approach is a planned order release mechanism in which orders are placed at regular intervals with a timing independent of the rate of demand (as opposed to reorder point system).
2. All parts have a common lead time and all orders in a specific cycle have the same due date.
3. There is only one order release to a cell, resulting in less paperwork.
4. Work-in-process and component-parts inventories are reduced.
5. Direct material costs are reduced and common raw material may be cut, thus reducing scrap and obtaining maximum material usage.
6. The use of a short planning period enables the system to react rapidly to changes in market demand.

One of the main drawbacks of PBC is the absence of clear guidelines for determining the correct cycle length. Some attempts have been made to determine the optimal cycle length based on expected costs consisting of inventory holding and overtime incurred in satisfying the demand for all end-items (Kaku and Krajewski, 1995). Models for both general flowline-type fabrication and assembly cells have been developed.

9.3 OPERATIONS ALLOCATION IN A CELL WITH NEGLIGIBLE SETUP TIME

Due to the characteristics of cellular manufacturing systems, production planning problems differ from those of traditional production systems. Some of these characteristics are:

- the use of group tooling considerably reduces setup time;
- machines are more flexible in performing various operations;
- a large variety of parts, most with low demand;
- fewer machines than part types;
- the operation of cells using just-in-time manufacturing concepts makes the production planning horizon very short.

These characteristics alter the nature of production planning problems in GT/cellular manufacturing systems. For example, cellular manufacturing permits flexibility, i.e. an operation on a part can be performed on alternate machines. Consequently, it may take more processing time on a machine at less operating cost compared with less processing time at higher operating cost on another machine. Therefore, the allocation of operations for a minimum-cost production plan will differ from production plans for minimum processing time or balancing of workloads. In this section, simple mathematical programming models for operations allocation which meet these different criteria are presented.

Consider a cell with M machine types ($m = 1, 2, \ldots, M$), each with a capacity of b_m, in which K part types ($k = 1, 2, \ldots, K$) with demand d_k are manufactured. Assume that $J_k = (j = 1, 2, \ldots, J_k)$ operations are performed on part type k. The unit processing time and unit processing cost to perform an operation on a part are defined as follows:

$$C_{kjm} = \begin{cases} \text{unit processing cost to perform } j\text{th operation on } k\text{th part type} \\ \text{on machine } m \\ \infty, \quad \text{otherwise} \end{cases}$$

$$t_{kjm} = \begin{cases} \text{unit processing time to perform } j\text{th operation on } k\text{th part type} \\ \text{on machine } m \\ \infty, \quad \text{otherwise} \end{cases}$$

With more flexible machines, an operation can be performed on different machines; as a result, a part can be manufactured along several processing routes. For example, if there are three operations on a part and the first, second and third operation can be performed on 2,3 and 2 alternative machines, respectively, a set of alternate processing routes L would include $2 \times 3 \times 2 = 12$ different processing routes. To define a process route $l \in L$ the following coefficient is used:

$$a_{kljm} = \begin{cases} 1, \text{ if } j\text{th operation on } k\text{th part type is performed} \\ \text{on } m\text{th machine using plan } l \\ 0, \quad \text{otherwise.} \end{cases}$$

Let x_{kl} be the decision variable representing the number of parts of type k to be processed using plan (process route) l. Models to find the objective functions minimizing the total processing cost and total processing time to manufacture all parts are presented below. By allowing operations on parts to be performed on alternate machines, some machines will be more heavily loaded than others. Balancing the workload on machines in a cell is another important objective which can be achieved by minimizing the maximum workloads (processing times) on the machines. A model for balancing workloads is also presented.

Minimum total processing cost

Minimize

$$z_1 = \sum_{kljm} a_{kljm} c_{kljm} x_{kl} \tag{9.6}$$

subject to:

$$\sum_{l} x_{kl} \geqslant d_k, \quad \forall k \tag{9.7}$$

$$\sum_{klj} a_{kljm} t_{kljm} x_{kl} \leqslant b_m, \quad \forall m \tag{9.8}$$

$$x_{kl} \geqslant 0, \quad \forall k,l \tag{9.9}$$

In this model, constraint 9.7 indicates that the demand for all parts must be met; constraint 9.8 indicates that the capacity of machines should not be violated, and constraint 9.9 represents the non-negativity of the decision variables.

Minimum total processing time

Minimize

$$z_2 = \sum_{kljm} a_{kljm} t_{kljm} x_{kl}$$

subject to

$$\sum_{t} x_{kl} \geqslant d_k, \quad \forall k \tag{9.10}$$

$$\sum_{klj} a_{kljm} t_{kljm} x_{kl} \leqslant b_m, \quad \forall m \tag{9.11}$$

$$x_{kl} \geqslant 0, \quad \forall k,l \tag{9.12}$$

Balancing of workloads

Minimize z_3 subject to:

$$z_3 - \sum_{kljm} a_{kljm} t_{kljm} x_{kl} \geqslant 0, \tag{9.13}$$

$$\sum_{t} x_{kl} \geqslant d_k, \quad \forall k \tag{9.14}$$

$$\sum_{klj} a_{kljm} t_{kljm} x_{kl} \leqslant b_m, \quad \forall m \tag{9.15}$$

$$x_{kl} \geqslant 0, \quad \forall k,l \tag{9.16}$$

Example 9.6

Consider the manufacture of five part types on four types of machines. All information regarding part demands, available machine capacity, unit processing cost and processing time on each machine for each route is given in Table 9.11. Develop production plans using the minimum processing cost model, minimum processing time model and balancing of workloads model.

Table 9.11 Data for Example 9.6. (Source (also for Tables 9.12 and 9.13): *Singh, N. Systems Approach to Computer Integrated Design and Manufacturing*, 1996. Reproduced with permission from John Wiley & Sons, Inc., New York.)

	Machine types				
Parts/operation	*m1*	*m2*	*m3*	*m4*	*Demand*
1 Op1	(10,20)		(6,5,70)		100
Op2		(9.5,40)	(4.5,70)		
2 Op1			(6,80)	(10,60)	80
Op2	(7.5,60)		(6.5,70)		
Op3				(8.5,20)	
3 Op1		(13.5,10)	(8.5,25)		70
Op2	(9,40)		(5,25)		
4 Op1	(7,35)		(5.5,60)		50
Op2		(8.5,40)	(4,80)		
Op3		(11,10)		(9.5,20)	
5 Op1		(9.5,40)	(7,60)		40
Op2	(10.5,25)				
Capacity of machines	2400	1960	960	1920	

Blank entries mean that an operation cannot be performed on that machine.

Table 9.12 Production plan for various criteria

Parts		*Minimum cost production plan*	*Minimum time production plan*	*Production plan with workload balancing*
1	ml–m2	100	5	63
	ml–m3		95	
	m3–m3			37
2	m4–ml–m4	80	80	80
3	m3–m3	58		
	m2–m3	12	66	70
	m1–m2		4	
4	m1–m2–m2	4		
	m1–m2–m4	46		
	m1–m3–m2		4	4
	m1–m3–m4		46	46
5	m2–m1	40	40	40

Using the LINDO package to solve the three models, the results obtained are presented in Table 9.12. As can be seen more than one process plan can be used for the production of any part. Table 9.13 provides an insight into the resource utilization of various operation

Table 9.13 Machine loading for various operations allocation strategies

Machine types	*Minimum cost production plan*	*Minimun processing time production plan*	*Production plan with balancing of workloads*
m1	2400.00	2400.00	2045.00
m2	1960.00	1960.00	1960.00
m3	960.00	960.00	960.00
m4	1920.00	1557.00	1920.00

allocation strategies. From the slack analysis, it can be seen that various allocation strategies result in different resource utilization of machines. For example, resource utilization of machine m1 for the minimum cost, minimum time and balancing of workloads strategies is 2400, 2400 and 2045 units of time respectively. All three strategies result in 100% utilization of machines m2 and m4, making these bottleneck machines. This information is helpful in scheduling production of parts as well as preventive maintenance of machines.

9.4 MINIMUM INVENTORY LOT-SIZING MODEL

The objective in a just-in-time (JIT) manufacturing environment is to minimize the total inventory. To realize the benefits of JIT in a cellular manufacturing environment, it is important to develop lot-sizing models that consider tooling similarities. This involves forming families of components which share a major setup. Of course, individual components may have setups of their own. However, inventory is considered as a waste and the objective is to reduce inventory to the lowest possible level. Inventory should be avoided unless required due to limited capacity. Below, a model due to Erenguc and Mercan (1990) and a heuristic solution scheme developed by Mercan and Erenguc (1993) are presented. In formulating the model, the following assumptions are made:

1. products are grouped into families, creating two types of setup: family (major) setup time S_i; individual (minor) setup time S_j;
2. the setup is considered through the capacity constraint;
3. no backlogging, zero replenishment lead time, and inventory at the begining and end of the planning horizon is zero.

The objective is to determine the production schedule for each item in each period that minimizes the total inventory cost subject to the demand requirements and the capacity limitations in each period. This problem, formulated as a optimization problem (Erenguc and Mercan, 1990), is given below.

Notation

a_j capacity absorption rate by each unit of product j
B_t available capacity in period t
d_{jt} demand for product j in period t
h_j inventory holding cost for product j
I_i index set of products in family i
I_t^i index set of products in family i in period t
m number of families
r number of products
r_i cardinality of $I_i, \sum_{i=1}^{m} r_i = r$
T number of periods in the planning horizon
x_{jt} number of units of product j to be produced in period t
y_{jt} ending inventory of product j in period t

For each $i \in \{1,2,\ldots,m\}$ and each $t \in \{1,2,\ldots,T\}$, let $x_t^i = \{x_{jt} : j \in I_i\}$. The joint setup time function G_{it} for each family in each period t is given by

$$G_{it}(x_t^i) = \begin{cases} 0, & \text{if } \sum_{j \in I_i} x_{jt} = 0 \\ S_i, & \text{if } \sum_{j \in I_i} x_{jt} > 0 \end{cases}$$

For each $j \in I, t$ and x_{jt}, the capacity absorption function V_{jt} is given by

$$V_{jt}(x_{jt}) = \begin{cases} 0, & \text{if } x_{jt} = 0 \\ s_j + a_j x_{jt}, & \text{if } x_{jt} > 0 \end{cases}$$

For each x_t^i, the capacity absorption function τ_{it} is given by

$$\tau_{it}(x_t^i) = G_{it}(x_t^i) + V_{jt}(x_{jt})$$

The problem can be formulated as follows:

Minimize

$$f(x,y) = \sum_{t=1}^{T} \sum_{j=1}^{r} h_j y_{jt}$$

$$y_{jt-1} + x_{jt} - y_{jt} = d_{jt}, \quad \forall j,t \tag{9.17}$$

The capacity constraint in each period can be defined as:

$$\sum_{i=1}^{m} \tau_{it}(x_t^i) = \sum_{i=1}^{m} \left[G_{it}(x_t^i) + \sum_{j \in I_i} V_{jt}(x_{jt}) \right] \leqslant B_t, \quad \forall t = 1,2,\ldots,T \tag{9.18}$$

$$0 \leqslant x_{jt} \leqslant U_{jt}, \quad \forall j,t \tag{9.19}$$

$$y_{jt} \geqslant 0, \quad \forall j,t \tag{9.20}$$

$$y_{jT} = 0, \quad \forall j \tag{9.21}$$

$$y_{j_o} = 0, \quad \forall j \tag{9.22}$$

Heuristic solution procedure

The heuristic procedure starts by developing the lowest-cost production schedule, ignoring the capacity constraint, and then adjusting this schedule to achieve capacity-feasible production batches. If the capacity in each period satisfies the demand for all items in that period, then the optimal schedule is to carry no inventory. However, such a production schedule may violate the capacity constraints in certain periods and therefore will not be feasible. The idea of this heuristic is to find a feasible production schedule with no capacity violations by shifting the production of certain items to earlier periods when there is excess capacity. The following is a brief description of this heuristic.

Let $x_{jt} = d_{jt}$ (production = demand for all periods). If there are no capacity violations, this will be the best solution and terminates the procedure. But if there is a deficiency (capacity violation):

1. go to the largest period index t' where there is a deficiency;
2. set $x_{jt} = d_{jt}$ for all periods greater than t';
3. generate shift alternatives to shift a part or all of the lot-sizes of some products from period t' to period $(t' - 1)$.

Let $k(t')$ be the number of shift alternatives generated in period t'. Associated with each shift alternative is the partial (feasible) schedule V. Let $x(v,t')$ be the vth partial (feasible) schedule in period t'. For each shift alternative, the following need to be determined.

1. The partial (feasible) schedule for all periods greater than t': $x_{jt}(v) = d_{jt}$, but for period $t', x_{jt'}(v) = d_{jt'}(v) - z_{jt'}$, where $z_{jt'}$ is the amount of product j shifted to period $(t' - 1)$.
2. The trial production quantity for period $(t' - 1)$:

$$q_{j(t'-1)(v)} = d_{j(t'-1)(v)} + Z_{jt'}$$

3. The cost of holding inventory $C[(v,t'),(p,t' - 1)]$, which represents the cost of holding the shifted parts plus the cost of the pth shift alternative from which the v shift in period t' is generated. This cost is calculated for each $K(t)$ alternative and is ranked in increasing order. The first $k(t) \leqslant K(t)$ is then chosen.

For any v (shift alternative) for which the trial quantities $q_{j(t'-1)(v)}$ violate the capacity constraint in period $t' - 1$, a number of shift alternatives which eliminate the infeasibility in period $t' - 1$ are generated. This process is repeated until $t = 2$. The complete feasible schedule $x(v,1) = \{x_{jt(v)}: j \in I, t \in \{1,2,\ldots,T\}\}$ with the minimum cost $C[(v,2),(p,3)]$

among the $K(2)$ schedules is chosen as the 'best solution' generated by this heuristic.

With this understanding, the heuristic procedure can now be formally stated.

Step 0. Initialization.

(a) Set $t' = 0$.
(b) Determine $K(t)$ for all $t = \{1,2,\ldots,T\}$.
(c) Set $q_{jt} = d_{jt}$ and determine the largest period index t' where there is a deficiency.

(i) If $t' = 0$, set all $x_{jt} = d_{jt}$ and terminate the procedure.
(ii) If $t' = T$, set:

$$K(t'+1) = k(t'+1) = 1$$

$$x_{jt'(1)} = d_{jt'} = x'_{jt'}$$

$$q_{jt'(1)} = d_{jt'} = x'_{jt'}$$

$$C[(1,t'),(1,t'+1)] = 0$$

(iii) If $t' < T$, for all $j \in I$ and $t' < t < T$:

$$K(t) = k(t) = 1.$$

$$x_{jt}(1) = d_{jt} = x'_{jt}$$

$$q_{jt}(1) = d_{jt} = x'_{jt}$$

$$C[(1,t),(1,t+1)] = 0$$

(iv) Set $t = t'$.

Step 1. Generate shift alternatives for each of the $k(t+1)$ alternatives where the trial quantities $q_{jt}(v)$ violate the capacity constraints.
Step 2. For each $v = \{1,2,\ldots K(t)\}$, determine the feasible partial schedule $x(v,t) = \{x_{js(v)}{:}s = \{t,t+1,\ldots,T\}\}$ by subtracting the units shifted to period $t-1$ from $q_{jt}(v)$.
Step 3. Determine the cost $C[v,t),(p,t+1)]$ of each partial schedule. This represents the cost of carrying inventory of the items shifted to period $t-1$ plus the cost of the partial schedule $x(p,t+1)$ from which $x(v,t)$ follows.
Step 4. Rank the $K(t)$ alternatives in increasing order of their cost values, $C[(v,t),(p,t+1)]$. Select the first $k(t) \leqslant K(t)$ of these alternatives.
Step 5. Complete the trial quantities $q_{j(t-1)(v)}$ by adding the number of units shifted (from period t to $t-1$) to $d_{j(t-1)}$ under each of the vth shift alternatives.
Step 6. Set $t = t-1$. If $t = 1$, go to step 7; otherwise, go to step 1.
Step 7. Set $x_{j1(v)} = q_{j1(v)}, \quad \forall j \in I$ and $v = \{1,2,\ldots,K(t)\}$. Select the complete feasible schedule $x(v,1) = \{x_{jt(v)}; \quad \forall j \in I, t = \{1,2,\ldots T\}$ with the minimum cost $C[(v,2),(p,3)\}$ as the best heuristic solution.

Table 9.14 Data for Example 9.7 (a)

Item	*Demand*						
	Period 1	*Period 2*	*Period 3*	*Period 4*	*Individual setup*	a_j	h_j
1	53	8	72	68	10	3	5.19
2	25	88	35	85	14	2	4.14
3	0	198	34	0	6	1	3.28
4	12	138	108	101	12	2	3.76
5	4	88	39	42	18	4	3.14
6	22	46	83	10	25	4	3.41
Capacity	1196	1875	1090	1094			

Table 9.14 (b)

Family	*Items in family*	*Family setup time*
1	1,2,3	168
2	4,5,6	249

Example 9.7
The procedure is illustrated using data from Mercan and Erenguc (1993). Six items are grouped into two families: parts 1,2 and 3 form family 1, parts 4, 5 and 6 form family 2. The remaining data are given in Table 9.14. There are two types of release schemes in generating shift alternatives:

- individual item release scheme, in which each product is considered independent of its family and is shifted independently;
- family release scheme, in which the total production of all items in a set of families are shifted from period t to $t-1$.

Consider the individual shifting scheme. Set $x'_{jt} = d_{jt}$ for all $j \in I$, $t \in \{1,2,\ldots,T\}$. Then compute the ratio (h_j / a_j) for all items and arrange them in increasing order: $h_5/a_5 = 0.79$, $h_6/a_6 = 0.85$, $h_1/a_1 = 1.73$, $h_4/a_4 = 1.88$, $h_2/a_2 = 2.07$, $h_3/a_3 = 3.28$. There is a capacity violation in period 4; by using the capacity constraint equation the deficiency can be calculated. $((168 + 249) + (10 + 14 + 12 + 18 + 25) + (68 \times 3) + (85 \times 2) + (0 \times 1) + (101 \times 2) + (42 \times 4) + (10 \times 4)) - 1094 = 1280 - 1094 = 186$ units. Thus, a shift is needed that would save at least 186 units of capacity in period 4.

Let $\mu(w), w \in \{1,2,\ldots,r\}$, be the product index with the wth smallest h_j / a_j ratio. To generate the first shift alternative, start with $w = 1$ (the smallest ratio), $\mu(1) = 5$. The first item to be shifted (from period 4 to 3) is item 5. The capacity saving from such a shift is $168 + 18 = 186$ capacity units. Item 5 will be completely shifted, and because the deficiency is completely eliminated, no other item needed to be released. The following quantities should be computed after each shift:

$$q_{53}(1) = 39 + 42 = 81;\ q_{j3}(1) = x'_{j3}, \quad \forall j \in I, j \neq 5$$
$$x_{54}(1) = 0;\ x'_{j4}(1) = x'_{j4}, \quad \forall j \in I, j \neq 5$$

The cost of this shift is $C[(1,4),(1,5)] = 0.0 + (42 \times 3.14) = 131.88$. To generate the second shift alternative, since $w = 2$, start by shifting item 6 which will save $40 + 25 = 65$ capacity units. The remaining deficiency is 121 units. Item 5 is next to be shifted. By shifting $121/4 = 30.25$ units this deficiency will be eliminated. Then compute the following:

$$q_{63}(2) = 83 + 10 = 93\ ;\ q_{53}(2) = 39 + 30.25 = 69.25;\ q_{j3}(2) = x'_{j3}, \quad \forall j \neq 5$$

$$q_{64}(2) = 0.0;\ q_{54}(2) = 42 - 30.25 = 11.75;\ x_{j4}(2) = x'_{j4}, \quad \forall j \neq 5 \text{ and } 6$$

The cumulative cost of this alternative is $C[(2,4),(1,5)] = 0.0 + (10 \times 3.41) + (30.25 \times 3.14) =$ 129.93. After generating all possible shift alternatives and moving backwards until $t = 2$, the alternative with the lowest cost can be chosen as the best possible solution.

9.5 SUMMARY

Production planning and control is concerned with manufacturing the right product types in the right quantities at the right time at minimum cost while meeting quality standards. The market barriers are coming down; the market is now open to global competition. Further, the technical complexity of products is increasing; the market demands shorter product life-cycles, high quality and low cost. To compete in such a scenario, it is important to have an integrated manufacturing planning and control system that can exploit the similarities in a discrete product manufacturing environment.

This chapter has provided an understanding of the production planning process in a general manufacturing environment. A conceptual understanding of demand management, aggregate production planning, the master production schedule, rough-cut capacity planning, material requirements planning, detailed capacity planning, order release and shopfloor scheduling and control has been provided. These concepts were illustrated with numerical examples. We provided a production planning framework that integrates MRP and GT. The well known period batch control approach was covered. Some mathematical models that exploit the flexibility inherent in cellular manufacturing, such as group setup time and performing operations on alternate machines, were also given.

APPENDIX: Data file for Example 9.2

Min $100 \times 1 + 14W1 + 2001 + 50\ U1 + 3I1 + 400B1 + 14H1 + 30F1$
$100 \times 2 + 14W2 + 2002 + 50\ U2 + 3I2 + 400B2 + 14H2 + 30F2$
$100 \times 3 + 14W3 + 2003 + 50\ U3 + 3I3 + 400B3 + 14H3 + 30F3$
$100 \times 4 + 14W4 + 2004 + 50\ U4 + 3I4 + 400B4 + 14H4 + 30F4$
$100 \times 5 + 14W5 + 2005 + 50\ U5 + 3I5 + 400B5 + 14H5 + 30F5$
$100 \times 6 + 14W6 + 2006 + 50\ U6 + 3I6 + 400B6 + 14H6 + 30F6$

Subject to:

$X1 - I1 + B1 = 300$
$X2 + I1 - B1 - I2 + B2 = 400$
$X3 + I2 - B2 - I3 + B3 = 450$
$X4 + I3 - B3 - I4 + B4 = 410$
$X5 + I4 - B4 - I5 + B5 = 300$
$X6 + I5 - B5 - I6 + B6 = 240$
$-W1 + W0 + H1 - F1 = 0$
$-W2 + W1 + H2 - F2 = 0$
$-W3 + W2 + H3 - F3 = 0$
$-W4 + W3 + H4 - F4 = 0$
$-W5 + W4 + H5 - F5 = 0$
$-W6 + W5 + H6 - F6 = 0$
$01 - U1 - \times 1 + W1 = 0$
$02 - U2 - \times 2 + W2 = 0$
$03 - U3 - \times 3 + W3 = 0$
$04 - U4 - \times 4 + W4 = 0$
$05 - U5 - \times 5 + W5 = 0$
$06 - U6 - \times 6 + W6 = 0$
END

REFERENCES

Burbidge, J. L. (1975) *The Introduction of Group Technology*, Wiley, New York.

Bitran, G. R. and Hax, A. C. (1981) Dis-aggregation and resource allocation using convex knapsack problems with bounded variables. *Management Science*, **27** (4), 431–41.

Bitran, G. R. and Hax, A. C. (1977) On the design of hierarchical production planning systems. *Decision Sciences*, **8** (1), 28–54.

Bitran, G. R., Hass, E. A. and Hax, A. C. (1981) Hierarchical production planning: a single state system. *Operations Research*, **29** (4), 717–43.

Erenguc, S. and Mercan, H. M. (1990) A multi-family dynamic lot sizing with coordinated replenishments. *Naval Research Logistics*, **37**, 539–558.

Ham, I., Hitomi K, and Yoshida, T. (1985) *Group Technology*, Kluwer Nijhoff Publishing, Boston.

Hax A. C. and Candea, D. (1984) *Production and Inventory Management*, Prentice-Hall, Englewood Cliffs, NJ.

Hyer, L. and Wemmerlov, U. (1982) MRP/GT: A framework for production planning and control of cellular manufacturing. *Decision Science,* **13** (4) 681–701.

Johnson, L. A. and Montgomery, D. C. (1974) *Operations Research in Production Planning, Scheduling, and Inventory Control,* Wiley, New York.

Kaku, B. K. and Krajewski, L. J. (1995) Period batch control in group technology. *International Journal of Production Research,* **33**, 79–99.

Mercan, H. M. and Erenguc, S. S., (1993) A multi-family dynamic lot sizing with coordinated replenishments: a heuristic procedure. *International Journal of Production Research,* **37**, 173–89.

Naidu, M. M. and Singh, N. (1986) Lot sizing for material planning systems – an incremental cost approach. *International Journal of Production Research,* **24** (1), 223–40.

Singh, N. (1996) *Systems Approach to Computer-Integrated Design and Manufacturing,* Wiley, New York.

Wagner, H. and Whitin, T. (1958) Dynamic version of economic lotsize model. *Management Science,* **5**, 89–96.

FURTHER READING

Bedworth, D. D. and Bailey, J. E. (1987) *Introduction to Production Control Systems,* 2nd edn, Wiley, New York.

Bitran, G. R., Hass, E. A. and Hax, A. C. (1982) Hierarchical production planning: a two stage system. *Operations Research,* **30** (2), 232–51.

Collins, D. J. and Whipple, N. N. (1990) *Using bar code: why it's taking over,* Data Capture Institute, Dusbury, MA.

Hitomi, K. (1982) *Manufacturing Systems Engineering,* Taylor & Francis, London.

Naidu, M. M. and Singh, N. (1987) Further investigations on the performance of incremental cost approach for lot sizing for material requirements planning systems. *International Journal of Production Research,* **25** (8), 1241–6.

Rolstadas, A. (1987) Production planning in a cellular manufacturing environment. *Computers in Industry,* **8**, 151–6.

Singh, N., Aneja, Y. and Rana, S. P. (1992) A bicriterion framework for operations assignments and routing flexibility analysis in cellular manufacturing systems. *European Journal of Operational Research,* **60**, 200–10.

Vollman, T. E., Berry, W. L. and Whyback, D. C. (1984) *Manufacturing Planning and Control Systems,* Richard D. Irwin, Homewood, IL.

CHAPTER TEN

Control of cellular flexible manufacturing systems

Jeffrey S. Smith and*

Sanjay B. Joshi†

Earlier chapters described the techniques and tools available for the creation of flexible manufacturing cells and systems. A flexible manufacturing system is a collection of machines (CNC machine tools) and related processing equipment linked by automated material handling systems (robots, AGVs, conveyors etc.), typically under some form of computer control. This chapter focuses on the control aspect of such systems. At this stage it is assumed that the FMS design is completely specified, i.e. the family of parts to be produced has been determined, the machines and equipment required have been specified, tooling and fixturing requirements have been established and the layout is complete. The problem now in hand is to develop a control system that will take manufacturing plans and objectives and convert them into executable instructions for the various computers that will be used to control the system. The execution of the instructions at the various computers, and ultimately at the machines and equipment, results in the operation of the system and the production of goods.

The software that performs the execution of instructions is called the shop floor control system (SFCS). Shopfloor control implements or specifies the implementation of the manufacturing plan as determined by the manufacturing planning system (MRP, Kanban etc.). As such, the SFCS interacts with, and specifies, the individual operations of the equipment and the operators on the shopfloor. The SFCS also tracks the locations of all parts and moveable resources in real time, or according to some predefined time schedule. An input–output diagram of the general shopfloor control problem is shown in Fig. 10.1, in which the

*Texas A&M University.
†Pennsylvania State University.

SFCS takes input from the 'mid-level' planning system and makes the minute-to-minute decisions required to implement the plan. As such, the SFCS provides a direct interface between the planning system and the physical equipment and operators on the shopfloor.

10.1 CONTROL ARCHITECTURES

The control architecture describes the structure of the control system. An 'architecture' is defined by the American Heritage Dictionary (176) as 'a style and method of design and construction', or 'a design or orderly arrangement perceived by man'. In terms of manufacturing control systems, Biemans and Blonk (1986) stated that 'an architecture prescribes what a system is supposed to do, i.e. its observational behavior in terms of inputs, outputs, and how these are related with respect to their time-ordering and contents'. Dilts, Boyd and Wherms (1991) pointed out that 'the performance of the control architecture, given the complex and dynamic environment of automated manufacturing, can ultimately determine the viability of the automated manufacturing system'. Jones (1984) suggested that the term 'architecture' is often used in data processing to describe the set of fundamental assumptions which underlie a technology.

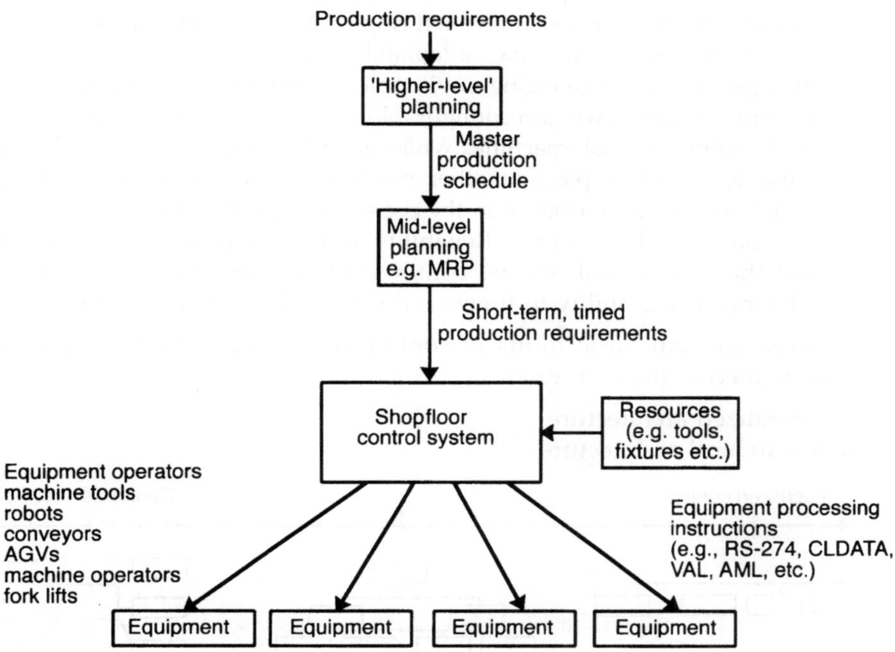

Fig. 10.1 Shopfloor control.

In the context of shopfloor control, a control architecture should provide a blueprint for the design and construction of a SFCS. It should completely and unambiguously describe the structure of the system as well as the relationships between the system inputs and outputs. Biemans and Vissers (1989, 1991) stated that 'abstract CIM architectures, describing control components in terms of their tasks and interactions, should form the basis for physical implementations of control systems'. In other words, the functionality of a system must be firmly established before the system can be implemented. This is certainly a requirement for generic and automatically generated systems. Furthermore, an architecture should depict a production organization in terms of a structure of interacting components which provides insight into how the components affect the behavior of the production organization as a whole. Dilts, Boyd and Whorms (1991) described the following demands which must be met by a control architecture in order for the developed SFCS to achieve technical and economic feasibility.

1. Modifiability/extensibility. Modifiability implies changes to the existing system can be easily made, whereas extensibility implies new elements can be easily added to the system to expand existing levels of functionality (note that these are design changes; reconfiguration required due to breakdowns is discussed next).
2. Reconfigurability/adaptability. Reconfigurability provides the ability to add or remove various manufacturing system components while the system is operational, and adaptability allows changes in control strategies based on changing environment conditions. For example, if a machine breaks down, a reconfigurable system will allow rerouting and rescheduling at that machine, while an adaptable system could also cause rerouting of parts at other machines to maintain and improve overall system performance in the presence of the machine failure.
3. Reliability/fault-tolerance. Reliability is the measure of probability that the system will operate continuously without failure, and fault tolerance is the ability to function despite failure of components.

There are four basic forms of control architecture which have been investigated in the literature:

- centralized architecture
- hierarchical architecture

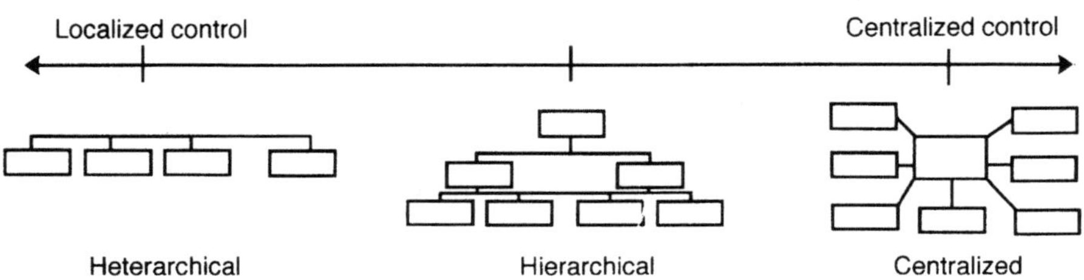

Fig. 10.2 Spectrum of control distribution (Duffie, Chitturi and Mou (1988)).

- heterarchical architecture
- hybrid architecture.

The distinction between these forms is in the interaction between the individual system components (Fig. 10.2). At the centralized control extreme, all decisions are made by a central controller and specific, detailed instructions are provided to the subordinate components. At the heterarchical extreme, on the other hand, the individual system components are completely autonomous and must cooperate in order to function properly. Each of these basic forms are described in more detail in the following sections and examples of each are provided.

Centralized control

Centralized control is one of the most common types of control for automated systems. Under this paradigm, a single workstation, mainframe, or minicomputer is connected directly to the equipment on the shopfloor. Figure 10.3 shows the structure of a centralized control architecture. A direct numerical control (DNC) system is a common example of a centralized control system. Often the control is implemented on a programmable logic controller (PLC) or other sequencing device.

The advantages of centralized control include:

- the centralized controller has complete access to global knowledge and information;
- overall system status can be retrieved from a single source;
- global optimization is easier to achieve.

The disadvantages include:

- reliance on a single central control unit, as a result, failure of the central unit will result in complete system failure;
- it is suitable only for relatively small systems; the speed of response gets slower as the system becomes large;
- modification/extension can be difficult.

Centralized control has been used extensively for FMSs. However, as these systems become larger and more complex, centralized control becomes more and more difficult. Distributing some or all of the control decisions is the answer for these systems. The following sections describe two different types of control distribution.

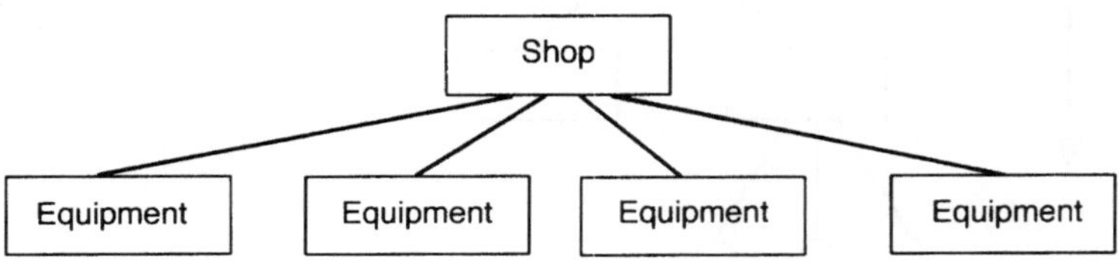

Fig. 10.3 Centralized control architecture.

Hierarchical models

In general, a hierarchical structure is used to manage the complexity of a system. Under the hierarchical paradigm, the functionality of the entire system is broken down into several levels in a tree-like structure (see Fig. 10.4). Each component in the hierarchical structure receives instructions from one immediate superior and provides instructions for several immediate subordinates. Using this approach, the size and complexity of any one component of the system can be limited to a manageable level. Warnecke and Scharf (1973) proposed the use of hireracbical control for integrated manufacturing. They stated the need for the following concepts to define an integrated manufacturing system:

1. a hierarchical framework;
2. product range flexibility with adaptive machines;
3. system integration using automated workpiece handling and tool changing;
4. enlargeability of the system;
5. compatibility with other systems.

These concepts have been the focus of much of the FMS and control architecture research described in this chapter.

The advantages of a hierarchical control architecture include:

- it provides a more modular structure as compared with the centralized approach;
- the modular structure allows gradual or incremental implementation; parts of the system can be made operational without the complete system being operational;
- the size, complexity and functionality of the individual modules is limited;
- the division of tasks within various levels allows for more natural partitioning and assignments of responsibilities;

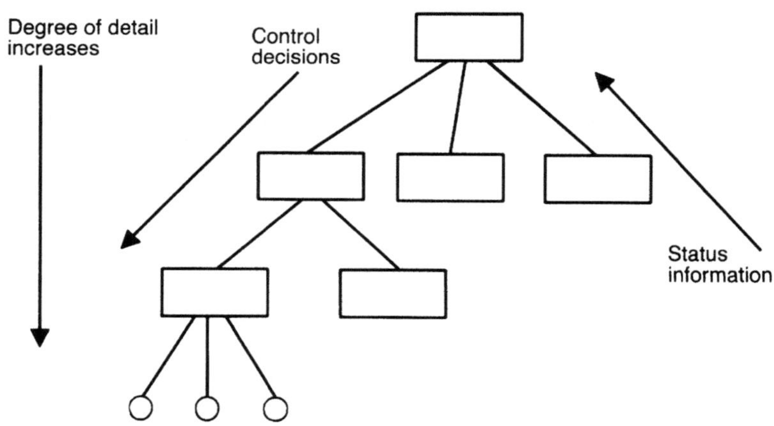

Fig. 10.4 Hierarchical control structure for shopfloor control.

- in the event of failure of a node in the hierarchy, only the branches below it would be affected, while the rest of the system may still be operational;
- since several computers are used in the hierarchy, in the event of a failure tasks may be shared by others.

Some disadvantages of the hierarchical structure include:

- increased software complexity, that is, while the complexity of any one module is controlled, there is significant overhead required to facilitate communication between modules;
- the need for aggregation and disaggregation of information, since controllers at different levels are performing at different frequencies;
- strict enforcement of the hierarchy creates long chains of command flow between controllers under different supervisors, which can lead to problems in reacting to real-time events;
- fault tolerance, although higher than centralized control, is lower than that of heterarchical control (described below).

Several hierarchical control architectures have been proposed and implemented over the last 10–15 years. The following sections discuss several notable implementations that have been described in the literature.

NBS/NIST control architecture

Albus, Barbera and Nagel (1981) described a hierarchical robot control system and identified three basic guidelines for developing manufacturing control hierarchies:

1. levels are introduced to reduce complexity and limit responsibility and authority;
2. each level has a distinct planning horizon and the length of this planning horizon decreases down the hierarchy;
3. control resides at the lowest possible level.

The robot control system introduces the concept of integrating hierarchically-decomposed commands from higher levels with status feedback from lower levels to generate real-time control actions (Fig. 10.5). This hierarchical control system forms the basis for the NIST hierarchical control architecture.

The NIST control hierarchy comprises five levels where each controller has one immediately higher level system controlling it and controls one or more systems in the level below it (Fig. 10.6). Each level in the hierarchy will combine the commands received from the higher level with the status feedback received from the lower levels to determine the required action. This action will then be performed by issuing commands to the immediately lower levels and providing status

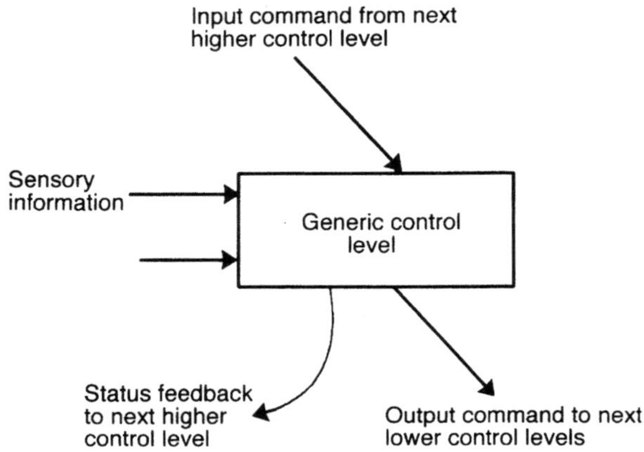

Fig. 10.5 Generic control level under hierarchical control.

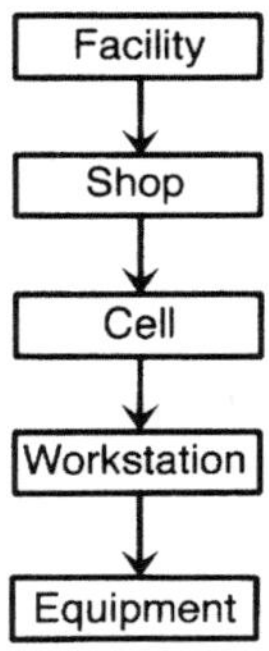

Fig. 10.6 NIST hierarchical control architecture.

feedback to the immediately higher level. The lowest level in the hierarchy (the equipment level) will implement the physical control of the equipment. Control is exercised using the 'state table' approach. A state table explicitly lists all possible system states and specifies an action to be taken when the system enters a particular state. When the action is performed, a 'state transition' takes place and the system enters another state. Once in the new state, the associated action is performed and another state transition occurs. The operation of the system can therefore be viewed as a sequence of states and state transitions. The state table model is described in more detail in section 10.3.

The 'facility' level is the highest level in the NIST hierarchy. It controls such long-range functions as cost estimation, inventory control, labor rate determination etc. The 'shop' level is responsible for coordinating activities between the manufacturing cells and allocating the required resources to the cells. The 'cell' level is responsible for sequencing batch jobs through the workstations and supervising the activities of the workstations. Materials handling between the individual workstations is a significant responsibility of the cell-level controllers.

The 'workstation' level controllers sequence and control the activities of the equipment controllers within each workstation. A typical workstation in the NIST control hierarchy consists of a material handling robot, one or two logically connected machine tools, and a material storage buffer. The workstation controller determines the physical tasks required for each operation assigned by the cell controller, sequences these tasks on the machines in the workstation, and coordinates the material handling via the robot. The 'equipment' level controllers are front-end computers for the machine tools and robots. They receive step-by-step commands from the workstation controller and convert them into the form required by the individual machine tools or robots. Smith (1990) presented a complete implementation of equipment- and workstation-level controllers for a FMS based on the NIST control hierarchy.

Manufacturing systems integration (MSI) control architecture

Recently, NIST has been working on revising and updating the original Automated Manufacturing Research Facility (AMRF) architecture through the Manufacturing System Integration (MSI) project (Senehi *et al.*, 1991). MSI addresses several of the shortcomings of the AMRF architecture which were discovered during implementation. Integration of systems is still the key issue that needs to be addressed. As a result, the emphasis of the MSI project is on the integration of manufacturing systems rather than on their development. The MSI architecture is similar to the original AMRF architecture in that it is hierarchical. However, the number of levels is not fixed but rather 'a level of control may be introduced whenever a coordinating or supervisory function is needed' (Senehi *et al.*, 1991). Introduction of new levels is seen as a system design activity rather than a dynamic activity. In other words, the hierarchical control configuration of the shop will not typically change once the system has been implemented (although the control hierarchy can be dynamically reconfigured to remove a dysfunctional piece of equipment (Senehi *et al.*, 1991)). The equipment level is the lowest level and is similar to the previous equipment-level definition. The shop level is the highest level in the hierarchy, and is also similar to the previous shop-level definition. However, between the equipment and shop levels exist a variable number of 'workcells'. A workcell coordinates the activities of two or more subordinate controllers, each of which is either an equipment or a workcell controller (Senehi *et al.*,

1991). Error recovery, process planning, human interfaces and global data management are all issues which are addressed in more detail than in the original architecture.

ESPRIT/CIM-OSA

The ESPRIT (European Strategic Programme for Research and Development in Information Technology) project was launched in 1984 as a 10-year program. The overall objectives of the ESPRIT project were (Macconaill, 1990):

1. to provide the European information technology (IT) industry with the basic technologies to meet the competitive challenge of the 1990s;
2. to promote European industrial cooperation in precompetitive research and development in IT;
3. to contribute to the development and implementation of international standards.

As part of this project, ESPRIT provides a comprehensive view of CIM. The emphasis of the ESPRIT strategy on CIM has been on developing standards and technology for multi-vendor systems. One of the primary outputs of the ESPRIT project has been CIM-OSA (Computer Integrated Manufacturing–Open Systems Architecture). A comprehensive description of CIM-OSA was presented by Beeckman (1989), Jorysz and Vernadat (1990a, 1990b) and Klittich (1990). CIM-OSA defines three main modeling levels (Beeckman, 1989):

1. enterprise model, describes in business terminology what needs to be done;
2. intermediate model, structures and optimizes the business and system constraints;
3 implementation model, specifies an integrated set of components necessary for effective realization of the enterprise operations.

These three models represent different stages in the building of the enterprise's physical CIM system. Similarly, each of the models is described in terms of four different views (Beeckman, 1989):

1. function view, the functional structure of the enterprise;
2. information view, the structure and content of information;
3. resource view, the description and organization of enterprise resources;
4. organization view, fixes the organizational structure of the enterprise.

CIM-OSA describes controllers from a system interaction viewpoint. Details of the operation of the individual controllers are not specified. Instead, CIM-OSA specifies how these controllers interface to external systems.

Heterarchical/agent-based models

Several researchers have expressed concern over the rigidity of the hierarchical structure. Hatvany (1985) pointed out the need for a new type of manufacturing control model which will:

- permit total system synthesis from imperfect and incomplete descriptions;
- be based on the automatic recognition and diagnosis of fault situations;
- incorporate automatic remedial action against all disturbances and adaptively maintain optimal operating conditions.

Hatvany (1985) suggested the application of the so-called 'law of metropolitan petty crime', which is described as the fragmentation of a system into small, completely autonomous units, each pursuing its own selfish goals according to its own self-made laws. The suggested application is in the form of cooperative heterarchies, or systems in which all participant subsystems should have:

- equal right of access to resources;
- equal mutual access and accessibility to each other;
- independent modes of operation;
- strict conformity to the protocol rules of the overall system.

Duffie, Chitturi and Mou (1988) pointed out that the organization and structure of hierarchical systems become fixed in the early stages of design and that extensions must be foreseen in advance, making subsequent unforeseen modifications difficult. They also proposed the use of a heterarchical control architecture and provided a detailed description. Conceptually, heterarchical systems are constructed without the master/slave relationships indicative of hierarchical systems. Instead, entities within the system 'cooperate' to pursue system goals. Elimination of global information is a major goal of heterarchical architectures and this elimination tends to enhance the following aspects (Duffie, Chitturi and Mou 1988):

- containment of faults within entities
- recovery from faults in other entities
- system modularity, modifiability and extendibility
- complexity reduction
- development cost reduction.

Duffie and Piper (1987) presented a part-oriented heterarchical control system in which each individual part and machine was represented by a system entity. Part entities have knowledge of the processing that they require and machine entities have knowledge of the processing that they can perform. Part entities 'broadcast' processing requirements over the system network and machine entities similarly broadcast processing availability. When a match is found, the part and machine entities negotiate and, once an agreement is made, the part is transported to the machine and processing begins. An implementation in which each entity is an 'intelligent' Pascal program running under a multitasking operating system is described. A software development cost saving of 89% over a similar hierarchical system was reported (based on the number of lines of code: 2450 lines versus 259 lines) (Duffie and Piper, 1987).

Upton, Barash and Matheson (1991) likened a heterarchical control system to the system that gets commuters to work. No global controller directs each vehicle, but the control objective is achieved through simple, distributed rules (i.e. each driver strives to minimize commuting time without regard for other drivers' objectives). Upton, Barash and Matheson (1991) presented some preliminary results on the use of heterarchical systems for manufacturing system control. The results are based on a simulated manufacturing system with a standard part flow and multiple possible machines (each with a different processing time for similar parts). They pointed out that, based on the simulations, the distributed architecture dispatches jobs as a centralized controller might (since there are no controller entities, the parts are not actually 'dispatched'; instead, they are accepted for processing and request transport on their own), using the best machines when idle and, progressively, the less effective machines when busier. Upton, Barash and Matheson (1991) stated that further research in process planning, communications and on-board information processing is required to make the heterarchical architecture feasible for shopfloor control.

Lin and Solberg (1992) presented a generic heterarchical framework for controlling the workflow in a computer-controlled manufacturing system. The framework is based on a market-like model and uses a combination of objective and price-based mechanisms. Under this system, the individual entities negotiate for services provided by other entities. Intelligent software agents act as the representative for each entity in the system. For example, the typical control system includes machine agents, part entity agent, pallet agents, fixture agents, shared buffer agents, AGV agents, tool agents etc. A job comes to the system with a set of processing requirements, a process plan, priority and an objective. The controlling and scheduling process will arrange the resources needed, including machines, tools, pallets, fixtures and transporters, to get a job done according to the processing requirements to satisfy the part objective, to coordinate the resource sharing of jobs in the system, and to manage the information flow within the coordinating process and the communications with other system components. Additional details of this system were provided by Lin and Solberg (1994).

The advantages of such heterarchical control systems include the following:

- Fault tolerance is high; if one component goes down, the other system components continue to operate largely unaffected.
- The ability to modify the cooperative decision-making protocols and methods allows for reconfigurability and adaptability.
- Minimizing the global information constraints the amount of information that must be transmitted between components.

The disadvantages of heterarchical control are the following:

- Maintaining local autonomy contradicts the objectives of optimizing overall system performance.
- Since individual operations are determined through negotiation, it is difficult (and often impossible) to predict the timing for each operation.

Hybrid architecture

The hybrid architecture exploits the advantages of both hierarchical and heterarchical control concepts. The master–slave relationship of hierarchical control is loosened, and the autonomy of components is increased. Entities operate under the control of a supervisor with limited cooperative capabilities. Such architectures are difficult to generalize and can take an infinite number of forms, depending on the specific installation. Table 10.1 summarizes the characteristics of the centralized, hierarchical and heterarchical architectures.

10.2 CONTROLLER STRUCTURE COMPONENTS

The remainder of this chapter describes the structure and development of a hierarchical cell control system. However, many of the concepts (especially the planning and control concepts) can be generalized to centralized, heterarchical and hybrid systems.

Among existing hierarchical architectures there is much debate over the required number of distinct levels. We identify three 'natural' levels (which are generalized from Joshi, Wysk and Jones (1990) and Jones and Saleh (1989)): from the bottom of the hierarchy to the top (as shown in

Table 10.1 Architectural characteristics

	Centralized	*Hierarchical*	*Heterarchical*
Modifiability/extensibility	Difficult	Moderate	Simple
Reconfigurability/adaptability	Moderate	Moderate	Simple
Reliability/fault tolerance	Low	Moderate	High
System performance	Global optimal possible	Global optimal possible, but difficult	Global optimal impossible

Fig. 10.7) are the equipment, workstation and shop levels. The equipment level is defined by the physical shopfloor equipment and there is a one-to-one correspondence between equipment-level controllers and shopfloor machines. The workstation level is defined by the layout of the equipment. Processing and storage machines that share the services of a material handling machine together form workstations. Finally, the shop level acts as a centralized control and interface point for the system.

Planning, scheduling and execution

As described by Joshi, Wysk and Jones (1990) and Jones and Saleh (1989), controller activities at each level in the hierarchy can be partitioned into planning activities, scheduling activities and execution activities: (The term 'execution' is used in place of the term 'control' as originally used by Joshi, Wysk and Jones (1990) distinguish it from control in the classical sense which encompasses execution and scheduling activities. Similarly, Jones and Saleh (1989) used the terms adaptation, optimization and regulation.) In this system, planning commits by selecting the controller tasks that are to be performed (e.g. planning involves selecting alternative routes and splitting part batches to meet capacity constraints). Scheduling involves setting start/finish times for the individual processing tasks at the controller's subordinate entities. Execution verifies the physical preconditions for scheduled tasks and subsequently carries out the dialogue with the subordinate controllers required physically to perform the tasks. Table 10.2 provides the typical planning, scheduling and execution activities associated with the equipment, workstation and shop levels in the control hierarchy. Figure 10.8 illustrates the flow of information/control within a controller during system operation.

Equipment level

Within the control hierarchy shown in Fig. 10.7, the equipment level represents a logical view of a machine and an equipment-level

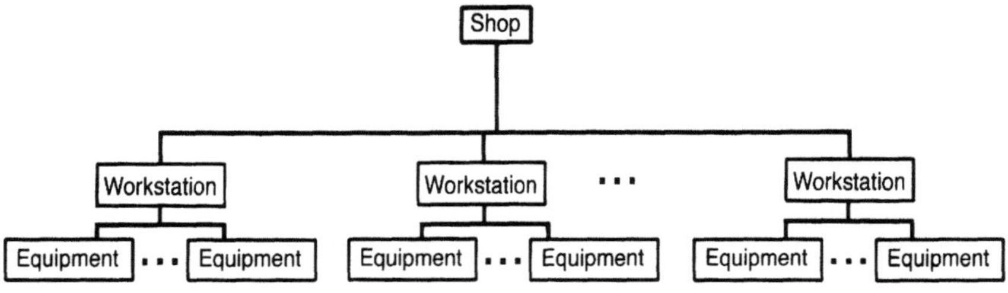

Fig. 10.7 Three-level hierarchical control architecture.

Table 10.2 Planning, scheduling, and execution activities for each level in the SFCS control architecture

Level	*Planning*	*Scheduling*	*Execution*
Equipment	Operations-level planning (e.g. tool path planning)	Determining the start/finish times for the individual tasks. Determining the sequence of part processing when multiple parts are allowed	Interacting with the machine controller to initiate and monitor part processing
Workstation	Determining the part routes through the workstation (e.g. selection of processing equipment). Includes replanning in response to machine breakdowns	Determining the start/finish times for each part on each processing machine in the workstation	Interacting with the equipment-level controller to assign and remove parts and to synchronize the activities of the devices (e.g. as required when using a robot to load a part on a machine tool)
Shop	Determining part routes through the shop. Splitting part orders into batches to match material transport and workstation capacity constraints	Determining the start/finish times for part batches at each workstation	Interacting with the workstation controllers and the resource manger to deliver/pick-up parts

controller. Informally we will refer to an equipment-level controller and its subordinate machine as simply a piece of equipment. Individual pieces of equipment also have machine controllers which provide physical control for the devices. These include CNC controllers, programmable controllers and other motion controllers, and are usually provided by the machine tool vendors. Equipment controllers provide a standard interface (based on the equipment type) to the rest of the control system. This interface hides the implementation-specific code required for machines from different vendors. An equipment-level controller makes decisions regarding local part sequencing, keeps track of part locations and monitors the operation of the machine under its control. Formally, the equipment level is defined as follows: $\mathbf{E} = \{e_1, e_2, \ldots, e_m\}$ is an indexed set of controllable equipment where

$$e_j \in \mathbf{E} \text{ and}$$
$$e_j = \langle EC_j, D_j \rangle$$

where EC_j is an equipment controller and D_j is a physical device (with

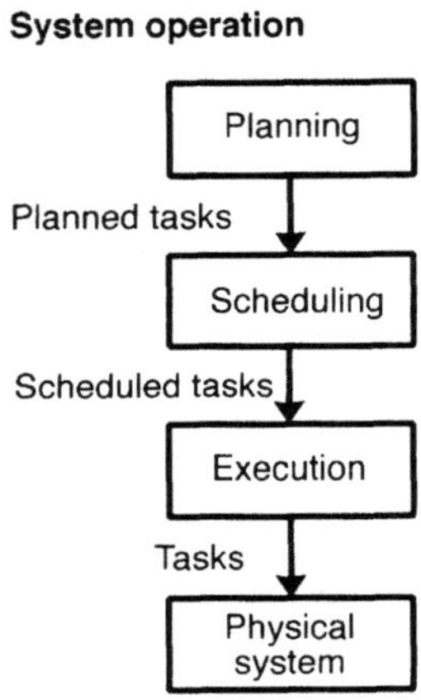

Fig. 10.8 Planning, scheduling and execution during system operation.

device controller). **E** is partitioned into {**MP, MH, MT, AS**} where:

$$\mathbf{MP} = \{e_j | D_j \text{ is a material processor}\};$$
$$\mathbf{MH} = \{e_j | D_j \text{ is a material handler}\};$$
$$\mathbf{MT} = \{e_j | D_j \text{ is a material transporter}\};$$
$$\mathbf{AS} = \{e_j | D_j \text{ is an automated storage device}\}.$$

The class of material processors includes machining centers, inspection devices, assembly machines, and so on. The key factor in placing a machine in this class is that the machine has the ability autonomously to 'process' a part in some way as indicated in the process plan. By 'processing,' we mean any activity that results in a change in the information content associated with the physical state of the part. For example, a turning center, a coordinate measuring machine and a painting booth are fundamentally different in terms of their processes, but from a control viewpoint each of these simply processes parts according to some set of instructions described by the process plan. A material processor may also have local storage and a dedicated load/unload device to move parts between the processing area and local storage. For example, many machining centers include a rotating index table or a pallet exchanger which hold multiple parts. These dedicated load/unload devices are controlled by the same device controller as the material processor. Based on the capability for local storage, each piece of equipment has a maximum capacity, indicating the maximum number of parts that can be assigned to that device at one time. Each unit of the capacity is designated as a location. A location can be addressable or non-addressable.An addressable location is reachable by an external device (e.g. a robot or an operator). For material processors, planning generally involves the development of numerical control programs (or their counterpart) for the individual parts. This includes tool selection and NC path planning. Scheduling involves determining the sequence of machining operations for each part.

The class of automated storage (AS) machines is made up of various AS/RS type devices. A piece of automated storage and retrieval machinery may store raw materials, work-in-process, finished parts, tools or fixtures. Objects are stored in locations known to the AS/RS controller. Storage machines can deliver any stored object to a load/unload point and can retrieve any object from a (possibly different) load/unload point and place it in storage. As with previous machines, automated storage machines have a capacity which consists of addressable and non-addressable locations. In general, the capacity of an automated storage device is much greater than the number of addressable locations. Planning for automated storage machines includes selecting storage locations for parts. Scheduling involves sequencing individual storage and retrieval tasks.

Machines used for moving objects within the manufacturing shop are separated into two classes. The class of material handling machines includes robots, indexing devices and other devices capable of moving parts from one location to another in a specified orientation. These locations are typically close together relative to the size of the factory. The primary function here is to load (unload) parts into (from) various material processors and automated storage machines. An individual piece of material handling machinery may have a capacity greater than one part by having multiple part-holding attachments (i.e. grippers).

The class of material transport machines is made up of AGVs, conveyors, fork trucks and other manual or automated transport machines. The primary function of these machines is to transport parts to various locations throughout the factory. The distinction between material handling machines and material transport machines is that the former handling machines can load and unload other equipment, and material transport machines cannot. Typically, material handling machines perform intra-workstation part movement functions and material transport machines perform inter-workstation part movement functions (workstations, as used here, are defined below). A specific type of material movement machine (e.g. a conveyor or a robot) could belong to either class, but within a particular system, each specific device (e.g. conveyor # 8 or a Puma robot) will be considered either a material handler or a material transporter, but not both. Associated with each material transport device is a set of 'ports'. A port is a location at which the individual transport device may stop to be loaded/unloaded. An example of a port is an AGV station where individual AGVs stop to be loaded and unloaded at a workstation. An individual port may be shared by several material transport devices. The set of all ports in the shop will be designated PO.

Each of the classes of machine defined above may include a part-holding device. In the case of a material processor this may be a chuck, vise or fixture. For material handling this will usually be a gripper. For material transport and automated storage and retrieval, parts may be held in pallets that act as vises or grippers. The concern here is that for a part to be removed from (or placed in) a work-holding device, an exchange of synchronization information may be required so that the part is not released by one device prior to being grasped by the other (which would create the potential for the part to be dropped).

Additionally, the following sets of non-controllable equipment, which require no machine controller are defined:

$$\text{BS} = \{\text{passive buffer storage units}\};$$
$$\text{PD} = \{\text{passive devices}\}.$$

The class of buffer storage units includes groups of passive storage locations to which a piece of material handling equipment has access. Buffer storage has a maximum capacity. A passive device is a special case of a material processor which requires no machine-level controller and has a deterministic processing time. An example of a passive device is a gravity-based device used to invert a part between successive turning processes to allow turning on both ends of the part. A buffer storage device is distinguished from a passive device by the fact that a passive device explicitly appears in the sequence of required operations for the part, and buffer storage is an optional operation performed between any two required operations.

Workstation level

A workstation is made up of one or more pieces of equipment under the control of a workstation-level controller. Workstations are defined using the physical layout of the equipment and are generally configured so that multiple MP devices can share the services of one or more MH devices and/or ports. We wish to create an indexed set of workstations, $\text{W} = \{\text{W}_1, \text{W}_2, \ldots, \text{W}_n\}$. To accomplish this, the sets MP, MH, MT, AS, BS and **PD** are each partitioned into subsets indexed by $i = 1, 2, \ldots, n$, corresponding to the indexing of W. For example, MP is partitioned into $\{\text{MP}_1, \text{MP}_2, \ldots, \text{MP}_n\}$. PO is defined as a finite set of ports. A port is a physical location at which parts can be transferred between pieces of equipment. **PO** is separated into (not necessarily disjoint) indexed sets $\text{PO}_1, \text{PO}_2, \ldots, \text{PO}_n$. A workstation W_i is then defined formally as:

$$\text{W}_i \in \text{W} \text{ and}$$
$$\text{W}_i = \langle \text{WC}_i, \mathbf{E}_i, \text{BS}_i, \text{PD}_i, \text{PO}_i \rangle,$$

where WC_i is a workstation controller and

$$\mathbf{E}_i = \{\text{MP}_i \cup \text{MH}_i \cup \text{MT}_i \cup \text{AS}_i\}.$$

The workstation controller carries out commands received from the shop controller and is responsible for moving parts between the various pieces of equipment in the workstation and for specifying part processing performed at this equipment. To this end, it will synchronize the actions required to coordinate the transfer of parts between processing equipment and material handling equipment. Since the individual equipment controllers are responsible for sequencing tasks once the tasks have been assigned by the workstation controller, the workstation is not responsible for loading, starting and monitoring the operation machine directly. Instead, parts are 'assigned' to the equipment controller, which specifies a 'delivery location' for the parts. Once the parts have been delivered, they are out of the direct control of the workstation. At some later time the equipment controller informs the workstation controller that the processing of the parts has been completed and provides a 'pickup location' for the parts. Between the delivery and pickup, the part is under the control of the subordinate equipment level controller.

Synchronization may also be required between the material handling equipment and a material transport device present at a port to deliver or remove parts. This would occur when parts are transported on fixtured pallets, for example. In this case the communication required for the synchronization will be with the shop controller rather than with the material transport device directly. The shop controller will, in turn, communicate with the transport workstation through the resource manager to facilitate the synchronization.

We identify three classes of workstation: processing, transport and storage workstations such that $W = \{W_P \cup EW_T \cup EW_S\}$. A processing workstation is a workstation that is made up one or more material processors MP and, optionally, one or more pieces of material handling equipment MH, one or more buffer storage devices (BS) or AS/RS. Formally,

$$W_P = \{W_i : |MP_i| > 0, |MT_i| = 0\}$$

An implicit assumption is that all addressable locations of the AS and BS devices and the ports within the workstation are accessible to at least one of the MP devices. This access is typically provided by a MH device which is used to load and unload the material processors. However, in some instances processing can be performed while the parts are at the port (e.g. a welding operation performed by a robot on parts moving on a conveyor line). Processing workstations perform all of the value-added processing required to transform raw materials into finished products.

Planning at the workstation level involves selecting the individual pieces of equipment at which the part will be processed. Workstation-level scheduling involves determining the part processing sequence within the workstation. When a part enters a processing workstation, it follows a 'workstation-level process plan'. This lists the various pieces of equipment in the workstation that the part must be sequenced across and the order in which the operations must occur. In the general case, a workstation-level process plan may include alternative routings for a part. Each of these routings can be viewed as a path through the workstation process-plan graph. A workstation-level process plan is a particular view of the equipment process-plan which includes only the equipment within the specific workstation.

A transport workstation is a workstation which is made up of one or more material transport devices (MT) and provides material transport services to the other workstations in the shop. A transport workstation might also include one or more MH devices for transferring parts from one MT device to another within the transport workstation. Formally,

$$W_T = \{W_i : |MP_i| = 0, |MT_i| > 0, |AS_i| = 0\}$$

The purpose of the transport workstation is to integrate (possibly) many different material transport devices into a single system so that the resource manager does not need to be concerned with which particular device will transport particular parts. Instead, the resource manager will simply request that objects to moved from a specified location to another specified location. Based on this request, the transport workstation will determine a set of feasible routes (each of which might contain multiple transport segments) to perform the move. The resource manager will then evaluate the alternatives and instruct the transport workstation on which move to perform. The use of a transport workstation will also localize the effects of the introduction of new or modified transport devices/systems on the control system.

Similarly, we define a storage workstation to integrate several material storage devices which are not assigned to particular processing workstations. The storage workstation might also include MH devices for loading (unloading) parts, tools, fixtures, etc. to (from) the storage device. Formally,

$$W_s = \{W_i : |MP_i| = 0, |MT_i| = 0, |AS_i| > 0\}$$

The storage workstation provides a centralized interface to a distributed storage system and, as with the transport workstation, will localize the effects of the introduction of new or modified storage devices on the control system.

Resource manager

The resource manager is a workstation-level entity which provides centralized access to shared resources. A shared resource is some resource that is used by several independent entities within the SFCS. It controls the storage and transport workstations and the tool and fixture management systems. Since the production requirements and part routes change frequently in the target environment, it is necessary to have transport capabilities between every pair of workstations within the shop. Similarly, it is important to have storage facilities to decouple the processing workstations. However, since these resources are shared, seizure of these resources by one workstation may affect other workstations in the shop. This also applies to the use of centralized tool and fixture management systems. Therefore, global knowledge is necessary to distribute or schedule access to these shared resources effectively. This is the job of the resource manager.

For example, consider the case where a new part is to be processed. The first step is to remove the raw materials from the storage workstation and transport them to the first processing workstation in the processing route. In the general case, the required raw materials could be stored in several storage facilities distributed throughout the facility. The transport times from each of these locations to the specified processing workstation will be different and will depend on the state of the transport system. Therefore, neither the storage workstation or the transport workstation alone has sufficient information to decide from which storage facility the part should be removed. This is the job of the resource manager. It receives the raw material locations from the storage workstation and the transport details from the transport workstation and makes a decision specifying a particular storage location and transport route. Identical situations exist in the transport of tools and fixtures.

The resource manager and its constituent workstations provide centralized access to (possibly) distributed resources which must be shared among many workstations. Owing to the complexity of the material transport task, it is expected that the material transport activities will be managed rather than scheduled. In this mode of operation, requests to the transport system are handled on a first-come first-served basis (although preemption is allowed), and the transport times are stochastic and are based on the current state of the transport system as a whole, that is, in terms of the traditional view of production scheduling, processing equipment is scheduled based on sequences of processes and their associated processing times. This is contrasted with the techniques used to dispatch material transport devices to service the shop once the schedule has been determined. The assumption is that there is an adequate capacity of transport equipment, and that this capacity has been managed at a level that can support any reasonable production schedule. Note that if the transport tasks could be scheduled (e.g. transport times could be accurately predicted a priori regardless of the state of the transport system), then the resource manager services could be scheduled directly by the shop controller. Formally,

$$RM = \langle M, W_T, W_s \rangle$$

where:

M = manager module (which includes the tool and fixture management systems)

$$|W_T| = 1,$$

and

$$0£^{1/2}W_s^{1/2}£1$$

Shop level

The 'shop' includes all workstations and the resource manager. The shop controller is responsible for selecting the part routes (at the workstation level), and for communicating with the resource manager for transport and storage services used to move parts, tools, fixtures etc. between workstations. The shop level is also the primary input point for orders and status requests and, therefore, has significant interaction with people and external computer systems. The shop level must also split part orders into individual batches to meet material transport and workstation capacity constraints.

Since all the components of the shop have been defined, a shop S can be formally defined as

$$S = \langle SC, W_p, RM \rangle$$

where:

SC is a shop controller. Furthermore, the following constraints are imposed on S:

$$|W_p| > 0$$

and

$$U\{PO_i | W_i \notin W_T\} = U\{PO_i | W_i \in W_T\}$$

The first constraint assures that there will be at least one processing workstation in the shop; the second relation assures that the ports in the processing and storage workstations are the same ports that are in the transport workstation (this assures that the processing and storage workstations are reachable by the equipment in the transport workstation). Figure 10.9 shows a layout and the corresponding formal description of the Penn State CIM Laboratory.

10.3 CONTROL MODELS

Given the formal description of a manufacturing system, the next step is to develop the control logic and the associated control software

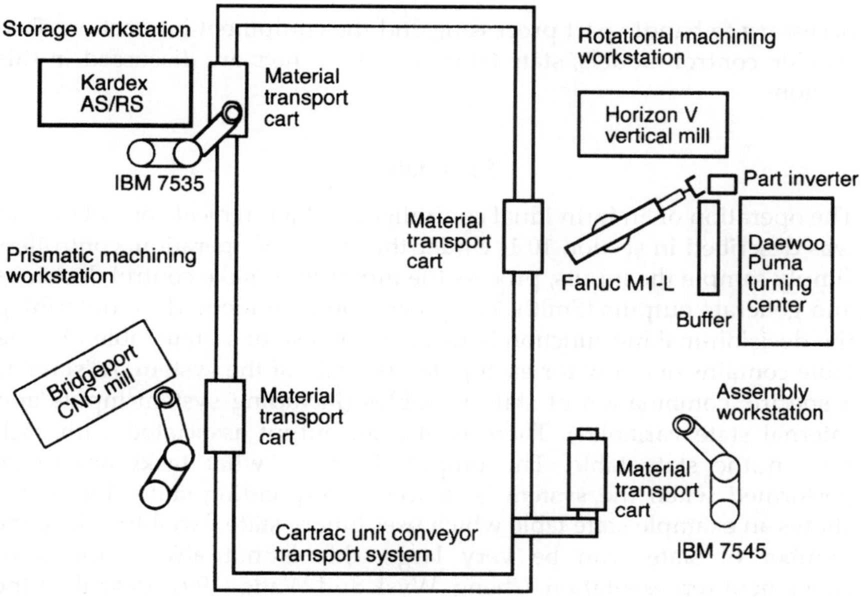

Shop level:
$S = <SC, \mathbf{W}_P, RM>$
$\mathbf{W}_P = (W_1, W_2, W_3)$
$RM = <M, \mathbf{W}_T, \mathbf{W}_S>$
$\mathbf{W}_T = (W_4)$
$\mathbf{W}_S = (W_5)$

Workstation level:
$\mathbf{W} = (W_1, W_2, W_3, W_4, W_5)$

$W_1 = <WC_1, \mathbf{E}_1, BS_1, PD_1, PO_1>$
$\mathbf{E}_1$ = (Puma, Horizon, Fanuc MI-L)
BS_1 = (Buffer)
PD_1 = (Part inverter)
PO_1 = (Cartrac 3)

$W_2 = <WC_2, \mathbf{E}_2, PO_2>$
$\mathbf{E}_2$ = (Bridgeport, Fanuc A0)
PO_2 = (Cartrac 4)

$W_3 = <WC_3, \mathbf{E}_3, PO_3>$
$\mathbf{E}_3$ = (IBM7545)
PO_3 = (Cartrac-4)

$W_4 = (WC4, \mathbf{E}_4, \mathbf{PO}_4)$
$\mathbf{E}_4$ = (SI Cartrac)
$\mathbf{PO}_4$ = (Cartrac-1, Cartrac-2, Cartrac-3, Cartrac-4)

$W_5 = (WC5, \mathbf{E}_S, PO_5)$
$\mathbf{E}_5$ = (Kardex, IBM7535)
$\mathbf{PO}_5$ = (Cartrac-2)

Equipment level:
E = (MP, MH, MT, AS)
MP = (Puma, Horizon, Bridgeport, IBM7545)
MH = (Fanuc MI-L, IBM7535, Fanuc AO)
MT = (SI Cartrac)
AS = (Kardex)
BS = (Buffer)
PD = (Part inverter)
PO = (Cartrac-1, Cartrac-2, Cartrac-3, Cartrac-4)

Fig. 10.9 Penn State CIM Laboratory.

necessary to handle part processing and the equipment interaction. Two specific control models, state tables and Petri nets are discussed in this section.

State tables

The operation of an individual controller in a hierarchical control system was described in section 10.1. Under this mode of operation, controllers simply sample the inputs, process the inputs and make control decisions and generate outputs (Smith, 1990). One common method for describing the decision-making function is through the use of a state table. A state table contains one row for each potential state of the system. A 'state' is a specific combination of state variables (including system inputs and internal state variables). There is also an output associated with each row in the state table. The output describes what tasks are to be performed when the system is in the corresponding state. Table 10.3 shows an example state table which uses binary state variables. Since the number of states can be very large, this is not always the most convenient representation. Chang, Wysk and Wang (1991) described the use of state table where the state variables are not binary. When the controller compiles the system state, it searches the state table for the corresponding entry. Once the table entry is found, the output associated with the state is performed.

Table 10.4 shows a state table for a small workstation containing a single robot used to load and unload a single machine tool. To reduce the size of the table, it is assumed that there is an infinite queue of parts waiting to be processed on the machine and an infinite output queue in which to place completed parts. The state table for this system contains eight rows corresponding to eight individual system states. The robot and machine state variables represent a part in contact with the device. The 'part complete' variable is required to distinguish between a new or in-process part on the machine (in which case there is no action required) and a completed part (in which case the robot should unload the part). Notice that three of these states are invalid. For example, state 2 represents the machine and robot being idle, but the part complete flag

Table 10.3 Example state table

State number	*System state definition*				*Output*
	y_2	y_3	$\cdots$	y_n	
1	0	0		0	O_1
2	0	0		1	O_2
$\vdots$					$\vdots$
m	1	1		1	O_m

Table 10.4 Example state table for a machine tended by a robot

State	*Machine*	*Robot*	*Part complete*	*Output action*
1	0	0	0	Pick part from input
2	0	0	1	Invalid state
3	0	1	0	Load part on machine
4	0	1	1	Put part on output conveyor
5	1	0	0	Wait
6	1	0	1	Pickup part from machine
7	1	1	0	Invalid state
8	1	1	1	Invalid state

being true. Similarly, states 7 and 8 are also impossible since the robot cannot pick up a part when the machine already has a part loaded. In the controller operation, the system would start in state 0 (no parts loaded). The corresponding output is to pick a part from the input queue. After completing this task, the system would transition to state 3. From state 3 the robot would load the part on the machine and the system would transition to state 5. In state 5, the controller waits for the machine to complete the processing of the part. Once the machine completes processing, the system transitions to state 6 and the robot is instructed to unload the part from the machine. From state 4, the robot puts the part in the output queue and the system returns to state 0 and begins the processing cycle again.

The state table control model is relatively simple to understand and implement for small systems. However, the number of system states grows very rapidly as the number of machines and/or the system buffer capacity increase. Smith (1990) provided a detailed description of a state table-based workstation controller. Rippey and Scott (1983) highlighted the following advantages of using a state table:

1. it serves as a form of documentation of the controller functions;
2. the state table (and therefore the functionality of the controller) is easily extendible by adding rows and columns to the state table;
3. it provides a structure for the development of control software.

More detailed descriptions and examples of state tables for specific systems were presented by Haynes *et al.* (1984), Mettala (1989) and Jones and McLean (1986).

Petri nets

Many researchers have successfully applied Petri net theory to the design, testing and development of manufacturing control systems. Petri nets are abstract, formal models which describe the flow of information or control in a system. The main attraction of Petri nets is that they are capable of modeling multi-conditional concurrent processes and conflicting processes in a straightforward manner (Peterson, 1981). Furthermore, Petri nets are easily implemented in computer code. Merabet (1985) used Petri nets directly to model the interactions of parts, robots and the machine tools within a manufacturing cell. He stated that the ability of Petri nets to model concurrency and conflict can be exploited to enforce deadlock freeness in concurrent processes and mutual exclusion for conflict resolution. The paper went into detail on a specific implementation in a cell containing a robot, two machine tools and a coordinate measuring machine.

Zhou, DiCesare and Desrochers (1989) described a method for developing a Petri net model of manufacturing systems while preserving the properties of safeness, liveness and reversibility: safeness indicates the absence of overflows in a manufacturing system (ensures that the machines have finite capacity); liveness ensures that all possible operations are reachable by some set of actions; and reversibility means that the system will be able to get back to the initial state. These properties are ensured by starting with a simplified net for which the properties can be proven, and successively expanding the net while following rules guaranteed to maintain the properties. The simplified nets can be easily expanded by expanding the nodes in the initial net into sub-nets which represent the detailed processes. So, in the simplified net, a workstation might be represented by one node and, in the expansion, the node is replaced by a sub-net which provides a realistic model of the operation of the workstation.

Valavanis (1990) also described this hierarchical modeling paradigm for modeling flexible manufacturing systems. However, he suggested that there are severe limitations associated with the use of existing classes of Petri nets for modeling FMSs. A new type of net called an extended Petri net was described. Extended Petri nets define multiple types of places and multiple classes of tokens and transitions. An example developed for a two-machine manufacturing cell was presented.

Kasturia, DiCesare and Desrochers (1988) described an application of colored Petri nets to cell control. A colored Petri net is a modified Petri net where a set of 'colors' is used to increase the modeling power of the net. The color of the tokens in the input places of a transition must be in the color set of the transition for it to be enabled. For example, if multiple part types might be present in a part buffer at the same time, assigning a different color to each part type will allow the controller to determine which process is required for each part. The changes in the part as it is being processed can be modeled by changing the color of the token representing the part as it moves through the net.

Bruno and Marchetto (1987) use an extended Petri net called a PROT net (Process translatable net) to develop a prototype of a manufacturing cell. The main feature of PROT nets is that the implementation of the processes and synchronizations can be generated from the net automatically. A major emphasis of the paper is the translation of PROT nets into Ada program structures which will provide a rapid prototype of the system.

10.4 SUMMARY

This chapter has described the basic structure of a shopfloor control system for a flexible manufacturing system. In this context, the SFCS is responsible for transforming the manufacturing plan into the executable instructions required to manufacture the parts. As such, it accepts input from mid-level planning and controls the operation of the equipment on the shopfloor. The control architecture concept was defined and three control architectures were discussed, and several examples of each from the literature were presented. A specific hierarchical control architecture was also described in detail. Within this control architecture, the shopfloor control functions are partitioned into planning, scheduling and execution tasks. Finally, the use of state tables and Petri nets for implementing shopfloor control were described.

REFERENCES

Albus, J., Barbera, A. and Nagel, N. (1981) Theory and practice of hierarchical control, in *Proceedings of the 23rd IEEE Computer Society International Conference*, Washington D.C., pp. 18–39.

Beeckman, D. (1989) CIM-OSA: Computer integrated manufacturing-open systems architecture. *International Journal of Computer Integrated Manufacturing*, **2** (2) 94–105.

Biemans, F. and Blonk, P. (1986) On the formal specification and verification of CIM architectures using LOTOS . *Computers in Industry*, **7**, 491–504.

Biemans, F. and Vissers, C.A. (1989) Reference model for manufacturing planning and control systems. *Journal of Manufacturing Systems*, **8** (1), 35–46.

Biemans, F. and Vissers, C.A. (1991) A systems theoretic view of computer integrated manufacturing. *International Journal of Production Research*, **29** (5), 947–66.

Bruno, G. and Marchetto, G. (1987) Process-translatable Petri nets for the rapid prototyping of process control systems. *IEEE Transactions on Software Engineering*, **12** (2), 346–57.

Chang, T.C., Wysk, R.A. and Wang, B. (1991) *Computer Aided Manufacturing,* Prentice-Hall, Englewood Cliffs, NJ.
Dilts, D.M., Boyd, N.P. and Whorms, H.H. (1991) The evolution of control architectures for automated manufacturing systems. *Journal of Manufacturing Systems,* **10** (1), 79–93.
Duffie, N.A., Chitturi, R. and Mou, J. (1988) Fault-tolerant heterarchical control of heterogeneous manufacturing system entities. *Journal of Manufacturing Systems,* **7** (4), 315–27.
Duffie, N.A. and Piper, R.S. Non-hierarchical control of a flexible manufacturing cell. *Robotics and Computer Integrated Manufacturing,* **3** (2), 175–9.
Hatvany, J. (1985) Intelligence and cooperation in heterarchic manufacturing systems. *Robotics and Computer Integrated Manufacturing,* **2** (2), 101–4.
Haynes, L.S., Barbera, A.J., Albus, J.S. *et al.* (1984) An application example of the NBS robot control system. *Robotics and Computer Integrated Manufacturing,* **1** (1), 81–95.
Jones, T.C. (1984) Reusability in programming: a survey of the state of the art. *IEEE Transactions on Software Engineering,* **10** (5).
Jones, A.T. and McLean, C.R. (1986) A proposed hierarchical control architecture for automated manufacturing systems. *Journal of Manufacturing Systems,* **5** (1), 15–25.
Jones, A. and Saleh, A (1989) A decentralized control architecture for computer integrated manufacturing systems, in *IEEE Symposium on Intelligent Control,* pp. 44–9.
Jorysz, H.R. and Vernadat, F.B. (1990a) CIM–OSA part 1: total enterprise modelling and function view. *International Journal of Computer Integrated Manufacturing,* **3** (3/4), 144–56.
Jorysz, H.R. and Vernadat, F.B. (1990b) CIM–OSA part 2: total enterprise modelling and function view. *International Journal of Computer Integrated Manufacturing,* **3** (3/4), 157–67.
Joshi, S.B., Wysk, R.A. and Jones, A. (1990) A scaleable architecture for CIM shop floor control, in *Proceedings of CIMCON '90,* (ed. A. Jones), National Institute of Standards and Technology, Gaithersburg, MD, pp. 21–33.
Kasturia, E., DiCesare, F. and Desrochers, A. (1988) Real time control of multilevel manufacturing systems using colored Petri nets, in *Proceedings of the 1988 International Conference on Robotics and Automation,* pp. 1114–19.
Klittich, M. (1990) CIM–OSA part 3: CIM–OSA integrating infrastructure – the operational basis for integrated manufacturing systems. *International Journal of Computer Integrated Manufacturing,* **3** (3/4), 168–80.
Lin, G.Y. and Solberg, J.J. (1992) Integrated shop floor control using autonomous agents. *IIE Transactions,* **24** (3), 57–71.
Lin, G.Y. and Solberg, J.J (1994) Autonomous control for open manufacturing systems, in *Computer Control of Flexible Manufacturing Systems: Research and Development,* (eds S. Joshi and J. Smith). Chapman & Hall, London, pp. 169–206.
Macconaill, P. (1990) Introduction to the ESPRIT programme. *International Journal of Computer Integrated Manufacturing,* **3** (3/4), 140–3.
Merabet, A.A. (1985) Synchronization of operations in a flexible manufacturing cell: the Petri net approach. *Journal of Manufacturing Systems,* **5** (3), 161–9.
Mettala, E.G. (1989) Automatic generation of control software in computer integrated manufacturing. Pennsylvania State Univ. Ph.D. thesis.
Peterson, J.L. (1981) *Petri Net Theory and the Modeling of Systems,* Prentice-Hall, Englewood Cliffs, NJ.
Rippey, W. and Scott, H. (1983) Real time control of a machining workstation, in *20th Numerical Control Society Conference,* Cincinnati, OH.

Senehi, M.K., Barkmeyer, E., Luce, M. *et al.* (1991) Manufacturing systems integration initial architecture document. NIST Interagency Rep. 4682, National Institute of Standards and Technology, Gaithersburg, MD.

Smith, J.S. (1990) Development of a hierarchical control model for a flexible manufacturing system. Pennsylvania State Univ. Master's thesis.

Upton, D.M., Barash, M.M. and Matheson, A.M. (1991) Architectures and auctions in manufacturing. *International Journal of Computer Integrated Manufacturing*, **4** (1), 23–33.

Valavanis, K.P. (1990) On the hierarchical modeling analysis and simulation of flexible manufacturing systems with extended Petri nets. *IEEE Transactions on Systems, Man, Cybernetics*, **20** (1), 94–110.

Warecke, H.J. and Scharf, P. (1973) Some criteria for the development of integrated manufacturing systems in *2nd International Conference on Developments in Production Systems*, Denmark.

Zhou, M., DiCeaser, F. and Desrochers, A. (1989) A top-down approach to systematic synthesis of Petri net models for manufacturing systems, in *Proceedings of the IEEE International Conference on Robotics and Automation*, pp. 534–9.

Index

Page numbers appearing in **bold** refer to algorithms.

MIX
Papier aus verantwortungsvollen Quellen
Paper from responsible sources
FSC® C105338

If you have any concerns about our products,
you can contact us on
ProductSafety@springernature.com

In case Publisher is established outside the EU,
the EU authorized representative is:
Springer Nature Customer Service Center GmbH
Europaplatz 3, 69115 Heidelberg, Germany

Printed by Libri Plureos GmbH
in Hamburg, Germany